Resource Recovery of Municipal Solid Wastes

Peter J. Knox, editor

K.P. Barney
L.J. Beetschen
H.A. Bell
A. Bos
M.J. Brady
T.E. Brna
R.J. Bryan
P.S. Canzano
Philippe Cazanave
D.H. Cleverly
K. Daugherty
R.A. Denison
F.A. Ferraro
Abe Finkelstein
J.A. Gaspar
B.K. Goodman
J.M. Harris
Floyd Hasselriis
M.R. Herron
R.L. Johnson
R.G. Kellam

J.D. Kilgroe
Takuya Kotani
R.E. Landreth
Anders Larkert
B.I. Loran
Katsuhiko Mikawa
P. Moore
P.J. Morris
R.M. Morrison
O. Ohlsson
M. Poslusny
B.L. Riddle
W.J. Sim
D.L. Sokol
R.E. Sommerland
L.A. Sowa
D.B. Spencer
C.R. Strong
C.E. Thomas
N.C. Vasuki
B. Venables

Donald K. Walter

AIChE Symposium Series
1988

Number 265

Volume 84

Published by
American Institute of Chemical Engineers

345 East 47 Street

New York, New York 10017

Copyright 1988

American Institute of Chemical Engineers
345 East 47 Street, New York, N.Y. 10017

AIChE shall not be responsible for statements or opinions advanced in papers or printed in its publications.

Library of Congress Cataloging-in-Publication Data
Resource recovery of municipal solid wastes.

(AIChE symposium series ; no. 265
Papers presented at the 1988 AIChE Spring National Meeting in New Orleans.
Includes index.
1. Salvage (Waste, etc.)—Congresses. 2. Resource recovery facilities—Congresses I. Knox, Peter J., 1949- . II. National Meeting of AIChE (1988 : New Orleans, La.). III. Series.
TD794.5.R4566 1988   628.4'458   88-8052
ISBN 0-8169-0454-5

Authorization to photocopy items for internal or personal use, or the internal or personal use of specific clients, is granted by AIChE for libraries and other users registered with the Copyright Clearance Center (CCC) Transactional Reporting Service, provided that the $2.00 fee per copy is paid directly to CCC, 21 Congress St., Salem, MA 01970. This consent does not extend to copying for general distribution, for advertising or promotional purposes, for inclusion in a publication, or for resale.

Articles published before 1978 are subject to the same copyright conditions and the fee is $2.00 for each article. AIChE Symposium Series fee code: 0065-8812/88 $2.00

Printed in the United States of America by
Twin Production & Design

FOREWORD

There is growing national emphasis on resource recovery and utilization of municipal solid wastes. The purpose of this book is to provide an overview of the current technology options, provide a discussion of the environmental/economic/political issues, and to focus attention on the relevance of chemical engineering and of chemical engineers in the solution of these problems.

Numerous books and articles have been published over the past 10 years on resource recovery. This book is unique however, because of its timeliness and its integration of the following elements:

1. The problems and successes of commercial size refuse-derived (RDF) plants in the United States and abroad.
2. Research and development efforts that are looking at the next generation of RDF technology options.
3. International perspectives (Jananese, Swedish, French, Italian and Dutch) on waste disposal practices.
4. Highlights of technical, environmental, ecomonic and political issues.
5. State-of-the-art landfill design considerations for protection of groundwater.

Most of the papers in this book were presented during four sessions at the 1988 AIChE Spring National Meeting in New Orleans. Special acknowledgements to the following people who helped with the development of the symposium are in order: N.C. Vasuki and Pat Canzano of the Delaware Solid Waste Authority, Gary Blasius of Occidental Chemical Company's Energy-From-Waste Plant, Dr. R.C. Knox of University of Oklahoma, and Dr. Nathan Snyder of the Ralph M. Parsons Company.

Special thanks to Engineering-Science, Inc. and The Ralph M. Parsons Company for their support in preparing this important book. Also, to the publications staff at AIChE (Maura Mullen and Maryanne Spencer) for their cooperation and patience.

> Peter J. Knox
> Manager of Industrial, Hazardous &
> Solid Waste Engineering Department,
> Engineering-Science, Inc.,
> Pasadena, California
>
> and
>
> 1987-1988 National Chairman of
> AIChE Solid Fuels and Alternate
> Energy Subcommittee (F & PD)

INTRODUCTION

In the early 1970's resource recovery from municipal solid wastes received much attention in the United States. This was primarily due to the energy crisis which gave impetus to the development of alternative fuels. Another reason was the search for an alternative to problems that have been encountered with the mass-burn technologies. In the aftermath of that period, a number of commercial size refuse-derived fuel (RDF) plants were built. Before they became operational, however, most of these plants required extensive modifications and redesign which resulted in a trend back to the use of mass-burn systems. Section 1 provides a discussion of the successes, failures, and lessons that have been learned at six different commercial size RDF plants in the United States.

Resource recovery from municipal solid wastes has gained increased attention in recent years due to the closing of numerous landfills. The difficulty in siting new landfills as a result of the NIMBY ("not in my back yard") syndrome is also another factor. Consequently, debate over the advantages and disadvantages of RDF versus mass-burn systems has been reawakened and the lessons learned at existing RDF plants provide a new benchmark for comparison.

The technical capabilities for solving our waste disposal problems already exist and this book provides a variety of examples of the different types of processing systems. The more fundamental long term problems, however, are political and cultural. Section 2 provides examples of different technologies developed in six different overseas countries, with insights into the cultural and political circumstances that have contributed to the success of these systems.

The success of an RDF or mass-burn facility depends on many factors other than the choice of a given technology. The plant size, the political and regulatory environment, and the bottom line economics are all factors to be considered. Ongoing research and development efforts will further increase the choice of options available in the future and Section 3 provides information on the next generation of RDF technologies involving thermal, biological, and densified RDF systems.

Environmental and economic considerations in processing municipal solid wastes are examined in Section 4. This includes an outline of the environmental consequences of various municipal solid waste disposal practices, characterization of refuse-derived fuels, state-of-the-art flue gas cleaning technologies, regulatory aspects of dioxin emissions, ash management issues, optimization of plant operations to minimize environmental risks, and trends associated with the financing of public works projects and their relationship to municipal solid waste projects.

Long term solutions to our solid waste disposal problems will involve a combination of different technologies depending on the locations and the political and cultural environments at those locations. This book provides the reader with a variety of options that are available to choose from. All of these options will still include the use of landfills and Section 5 provides insights into state-of-the-art sanitary landfill designs.

Peter J. Knox

CONTENTS

FOREWORD iii

INTRODUCTION iv

SECTION I: COMMERCIAL REFUSE DERIVED FUEL PLANTS IN THE UNITED STATES: PROBLEMS AND OPPORTUNITIES

STATE OF THE ART MATERIALS RECOVERY SYSTEMS OPTIONS AND ECONOMICS
.... N.C. Vasuki and Pasquale S. Canzano   1

THE AKRON, OHIO'S RECYCLE ENERGY SYSTEM: AN RDF SUCCESS STORY
.... David B. Spencer, Linda A. Sowa and Robert L. Johnson   14

METRO DADE COUNTY RESOURCES RECOVERY RDF FACILITY: PAST, PRESENT AND FUTURE .... Charles R. Strong   21

COLUMBUS, OHIO'S REFUSE DERIVED FUEL EXPERIENCE .... Henry A. Bell   25

REFUSE-DERIVED FUEL VERSUS MASS-BURN TECHNOLOGY: OGDEN MARTIN'S
 HAVERHILL/LAWRENCE, MASSACHUSETTS PROJECTS .... David L. Sokol   30

RESULTS OF EMISSIONS AND ASH TESTING AT A MODERN REFUSE-DERIVED FUEL PLANT .... Francis A. Ferraro   36

SECTION II: INTERNATIONAL DEVELOPMENTS IN THE PROCESSING OF MUNICIPAL SOLID WASTES

COMBUSTIBLE MATERIALS RECOVERY AND MSW FLUIDIZED BED INCINERATION IN JAPAN
.... Takuya Kotani and Katsuhiko Mikawa   44

AN ITALIAN PROCESS FOR HANDLING MUNICIPAL SOLID WASTES .... William J. Sim   55

PELLETIZING FOR WASTE UTILIZATION IN SWEDEN .... Anders Larkert   63

THE VALORGA SOLID WASTE REDUCTION PROCESS FROM FRANCE .... Lee J. Beetschen and Philippe Cazanave   71

WASTE MANAGEMENT IN HOLLAND AND THE LARGEST INTEGRATED WASTE
 PROCESSING PLANT IN THE WORLD .... Arian Bos   75

SECTION III: ADVANCED RDF TECHNOLOGIES

THERMAL AND BIOLOGICAL OPTIONS FOR ADVANCED RDF SYSTEMS
.... Donald K. Walter, Barbara J. Goodman and Christina E. Thomas   81

ORGANIC EMISSION STUDIES OF FULL-SCALE COFIRING OF PELLETIZED RDF/COAL
.... M. Poslusny, P. Moore, K. Daugherty, O. Ohlsson and B. Venables   94

SECTION IV: ENVIRONMENTAL AND ECONOMIC CONSIDERATIONS IN THE PROCESSING OF MUNICIPAL SOLID WASTES

ENVIRONMENTAL CONSEQUENCES OF MUNICIPAL SOLID WASTE DISPOSAL PRACTICES
.... Phillip J. Morris, Robert J. Bryan and Bruno I. Loran   107

ENVIRONMENTAL CHARACTERIZATION OF REFUSE DERIVED FUEL INCINERATOR TECHNOLOGY
.... Robert E. Sommerlad, Abe Finkelstein and James D. Kilgroe   115

STATE-OF-THE-ART FLUE GAS CLEANING TECHNOLOGIES FOR MSW COMBUSTION .... Theodore G. Brna   127

REGULATORY ANALYSIS OF POLLUTANT EMISSIONS, INCLUDING POLYCHLORINATED
 DIBENZO-P-DIOXINS (CDDs) AND DIBENZOFURANS (CDFs), FROM THE STACKS OF
 MUNICIPAL WASTE COMBUSTORS ..... David H. Cleverly, Rayburn M. Morrison, Brenda L. Riddle and Robert G. Kellam   141

THE HAZARDS OF MUNICIPAL INCINERATOR ASH AND FUNDAMENTAL OBJECTIVES
 OF ASH MANAGEMENT .... Richard A. Denison   148

HOW CONTROL OF COMBUSTION, EMISSIONS AND ASH RESIDUES FROM MUNICIPAL
SOLID WASTE CAN MINIMIZE ENVIRONMENTAL RISK .... Floyd Hasselriis 154

TRENDS IN FINANCING PUBLIC WORKS: EVALUATING PRIVATIZATION OF MSW DISPOSAL
FACILITIES .... Mark R. Herron and Kline P. Barney 168

SECTION V: PROTECTION OF GROUNDWATER AT SANITARY LANDFILLS

MANAGEMENT OF LEACHATE FROM SANITARY LANDFILLS .... Jeff M. Harris and James A. Gaspar 171

THE USE OF FLEXIBLE MEMBRANES IN PROTECTION OF GROUNDWATER .... Robert E. Landreth 185

A CASE STUDY OF SANITARY LANDFILL LEACHATE ATTENUATION BY A NATURAL
CLAY-PEAT SOIL IN THE GULF COAST REGION .... Michael J. Brady 192

STATE OF THE ART MATERIALS RECOVERY SYSTEMS OPTIONS AND ECONOMICS

N.C. Vasuki and Pasquale S. Canzano ■ Delaware Solid Waste Authority, P.O. Box 455, Dover, DE 19903

Solid waste disposal is now drawing extraordinary national attention because landfills are being closed at a rapid rate and new ones are not being opened. The supply of solid waste is inexhaustible and without any limitations. It is there as long as we have life on earth. The United States occupies the premier position in solid waste production. This is understandable because of the high standard of living achieved in the United States and the enormous choice of goods and services available to its residents. Solid waste disposal methods have been hampered by political indecision and not by technical limitation.

A long range solid waste disposal system has four main components. They are:

1. Materials recovery
2. Organic stabilization
3. Energy recovery
4. Long-term storage

Delaware Solid Waste Authority, Dover, Delaware

MATERIALS RECOVERY

Materials recovery is generally practiced in three different methods. The first method is scavenging materials of value from the waste stream. Although this is practiced in other countries such as Mexico, Egypt, India, China, etc., scavenging is generally outlawed in most of the industrialized western countries. The second method is source separation which requires homeowners and commercial firms to separate certain materials for recycling purposes before it enters the waste stream. Source separation could be either voluntary or mandatory. The third method is the central processing plant where either the source separated materials (but collected in one container) or the entire waste stream is processed in a plant to recover materials of value. The materials content of municipal solid wastes varies slightly from region to region. The typical content for Delaware is shown in Table 1. The easily separable materials are newsprint, yard wastes, glass, ferrous and non-ferrous metals.

Source separation systems require extraordinary cooperation of the population. If the homeowners become disenchanted with the program, the system will fail. Stores and other commercial establishments will continue the source separation of materials when high market values make it attractive. If the market-place is over supplied, the price falls and the commercial establishments also lose interest in source separation. Generally, a goal of 25 or 50% is stated in programs requiring source separation. That is 25% - 50% of the components in the waste stream are to be removed for reuse elsewhere. Table No. 2 shows the annual estimated yield of the easily recyclable materials from a population of 1,000 persons in Delaware. If the entire population cooperates, the expected yield of source separated recyclable materials is shown in the Table. Such a table could be used to measure the success of a source separation program.

Table 3 shows the source separation efficiency for items (a) through (d) as 47.5%. In other words, about 47.5% of the population participated in the program or, even with 100% participation, the ability to separate materials was only 47.5%. The overall participation rate and separation efficiency may vary week to week.

For yard wastes, the separation efficiency was 80%! This is reasonable because lawn mowing and yard clean up are special chores and because the material is collected on a separate day of the week. Assuming that all of the recyclables were sold or given away, the overall reduction (for items (a) through (d)) in the solid waste stream is 13.5% or 1,010 tons per year. If yard wastes are also included, then the reduction is 26.8% or 2,010 tons per year. For each township or county, similar tables can be prepared if the waste composition is known. The per capita numbers used in Table 1 reflect a typical suburban locality in Delaware.

By assigning values to the recovered materials and computing the separate collection cost, it is possible to determine the viability of a source separation system. In most cases, source separation programs are highly subsidized and the true costs are blurred.

Source separation systems are generally very popular with elected officials, because they provide the necessary short-term respite. Even though these programs do not provide a complete solution to the community's solid waste disposal problem, they act as a stop gap measure. They help increase the public awareness of the solid waste disposal costs. Source separation programs must be followed quickly with either energy recovery or landfilling options. If a follow up decision is not made in a timely fashion, the source separation program starts losing steam and people will revert to their habit of disposing the waste in the garbage can.

Materials recovery can also be achieved through a central processing plant where the community's waste are separated into various fractions. Such a plant can also accommodate source separated materials which are collected in the same truck. These are known as MRFs (Materials Recovery Facility).

ORGANICS STABILIZATION

Organics stabilization generally takes three different routes. Windrow composting is a method which has been used for many years in many parts of the world including the United States. Windrow composting has a potential for fire and odor problems. Therefore, "in-vessel" composting is becoming more prevalent. In such composting systems, the solid waste and/or sewage sludge is stabilized aerobically and pasteurization temperatures are reached within such vessels. A French firm Valorga[1] offers another option which is currently being tested on a smaller scale. This system utilizes a select organic fraction of the solid waste for anaerobic digestion to release methane gas. After such digestion, the remaining solid waste becomes a highly stabilized humus like product.

Organics stabilization is an option for a certain fraction of the waste stream such as yard wastes, food scraps and leaves. In the northern portion of the U. S., leaves become a major solid waste problem during the fall months. Many communities have an active leaf collection program under which raked leaves are disposed in a compost pile somewhere in the community. With the increased cost of landfilling and energy recovery, some communities are also opting for separate collection of the yard waste for composting purposes. Such separate collection systems are feasible when there are either franchised collection areas or municipally directed collection systems. Separate collection for composting systems is not very successful when there is free and open competition for solid waste collection. According to Goldstein[2], there are 107 composting systems currently in operation in the U.S., 15 are under construction and 39 more are in planning.

ENERGY RECOVERY

Energy recovery is now becoming a very popular method for disposal of solid waste. Success of both refuse derived fuel and mass burning systems have resulted in elected officials looking favorably at energy recovery systems. Generally, electric power is produced, although some systems supply both steam and electric power. Depending on the price of the energy sold, the cost of disposal can substantially vary and quite often disposal costs are subsidized by local taxes.

Energy recovery generally uses four different methods. The large size waterwall mass burning systems are generally preferred in larger cities. The modular waterwall mass burning units are preferred in smaller cities. The prepared fuels or RDF systems are favored where materials recovery is an important local issue. The modular mass burning combustion units with waste heat recovery are also popular. La Roc[3] provides a good description of technologies currently available in the U.S.

LONG-TERM STORAGE OR LANDFILLING

The last and the most critical of the main components of a stable solid waste system is the long-term storage or landfilling. Absent adequate capacity for landfilling, the system becomes unstable and approaches a crisis state. This is exactly what is happening in the northeastern United States and along the west coast. The landfilling option is required because it is the last stop for not only the ash from energy recovery system but also the rejects from the materials recovery and organics stabilization systems. In addition, there are certain fractions of the solid waste which have neither energy value nor materials recovery value and, therefore, must be disposed of in a landfill.

Long-term storage or the landfilling option will continue to be the main disposal option in the United States for at least the next two decades. Approximately 90% of the nation's solid wastes are disposed in landfills and this practice will continue until more energy recovery systems are constructed. Because of concerns about emissions from waste-to-energy plants, finding locations for constructing such plants is becoming exceptionally difficult, costly and time consuming. While projects are going through the siting and permitting process, the landfill becomes the ultimate backstop. In this respect, perhaps the more intelligent national option is to pay immediate attention to the construction of modern landfills while awaiting decisions on materials recovery, organics stabilization and energy recovery. Even if those other options are used, properly designed and constructed landfills will always be useful for the community. A proposal for recycling landfills is under study in Delaware.[4]

THE DELAWARE EXAMPLE - STATE OF THE ART MATERIALS RECOVERY SYSTEM

The State of Delaware took an extraordinary step in 1970 to appropriate $1 million for research and development of solid waste disposal systems. The Hercules Company was initially retained for this purpose and the concept of a multi-materials recovery system, known as the Delaware Reclamation Plant (DRP) emerged.

Although Hercules had developed a preliminary design for the plant in 1973, construction of the plant was delayed.

SYSTEM DESCRIPTION

The DRP was completed in 1983. It is capable of processing daily 1,000 tons of municipal solid wastes and up to 350 tons of municipal sewage sludge (20% solids). This Plant is an integral portion of a comprehensive waste disposal program for New Castle County, Delaware (population 420,000).

This County generates 500,000 tons of solid wastes and 60,000 tons of sewage sludge (20% solids) annually. This total includes household, commercial and industrial solid wastes. Of the total, approximately 250,000 tons per year of mainly household and light commercial solid waste are processed through the DRP. About 50% of the solid waste processed is recovered as Refuse Derived Fuel (RDF) which is then combusted in an Energy Generating Facility (EGF) located adjacent to the DRP. In addition, up to 60,000 tons of commercial waste (predominantly cardboard and wood) can be combusted in the EGF which uses 130,000 tons of RDF per year from the DRP as the main fuel. Despite this resource recovery effort, annually about 250,000 tons of solid wastes-consisting of ash from the EGF, the residue from the DRP, excess sludge, and solid wastes which have neither combustion capability or materials recovery value - are sent to the Cherry Island Landfill. This 250 acre landfill site has a series of cells with leachate collection systems. The leachate collected from the landfill (average 20,000 gpd) is pumped into the New Castle County sewer system and then treated at the City of Wilmington Waste Water Treatment Plant. The Pigeon Point Landfill adjacent to the EGF contains approximately 4 million tons of solid waste which were deposited in the landfill between 1972 and 1985. The methane gas generated by this landfill is now recovered and flared. The use of the gas as a source of fuel for additional steam generation is under study. A schematic representation of the system is shown in Figure 1.

PLANT DESCRIPTION

Refuse Derived Fuel (RDF) Recovery

The first module, Solid Waste Processing Module is shown in Figure 2. It consists of a large tipping and storage area (240 feet by 156 feet). The tipping and storage building contains two infeed conveyors (Heil metal pan conveyors, 8 feet wide) which feed two Heil vertical shaft, 1,000 h.p. hammermills (Heil Model 92B). Each shredder is capable of shredding 70 tons of solid waste per hour. Only one shredder is used for processing while the other shredder is on standby. The Heil shredders have been designed with explosion venting and Fenwal Explosion Suppression System. The shredders are outside the main tipping and storage building. This will minimize any damage to the tipping and storage building or to the workers, should there be an unforeseen explosion of a magnitude larger than anticipated. Over 1 million tons of solid wastes have been processed, and only three explosions have been experienced - an extraordinary record in the solid waste processing industry.

The shredded solid wastes (nominal size 4") are conveyed in enclosed conveyors to the dry processing building where they are air classified. All of the RDF conveyors have dust pickup, fire detection, and sprinkler systems. The exhaust air streams from the conveyors pass through a cyclone and a baghouse.

The dry process building houses the two main air classifiers. These air classifiers, manufactured by Iowa Manufacturing Company, are rotating drums (inclined at 15 degrees) with lifter bars. Suction is applied at the upper end of the drum thereby separating the light fraction from the shredded solid wastes. The light fraction of RDF consists predominantly of light paper and plastics with some stray metal and glass. The production rate is 500 tons/day.

Ferrous Recovery

The heavy fraction from the air classifier falls into a chute at the lower end of the drum where it is transferred to a conveyor which feeds the ferrous separation system consisting of a Dings Magnet. A cross-conveyor removes the ferrous metals (picked up by the Dings Magnet) to the outside of the building to a Newell drum magnet which removes the light ferrous fraction from the heavy ferrous fraction. The plant has achieved a production rate of 50 tons/day of clean light ferrous fraction (less than 1% loose nonmetalic impurities). The heavy ferrous fraction has no market value at this time. It is either given away to scrap dealers or landfilled.

Glass Recovery

The glass removal and processing steps are shown schematically in Figures 3 and 4. The solid waste fraction, after removal of the ferrous materials, proceeds to the wet process building where it is screened in Trommel G-1 (9' diameter and 24' long manufactured by Triple S Dynamics Corp.). The trommel has staggered 3/4" size holes. The trommel undersize, consisting predominantly of glass, is removed and stored in a bin (2,700 cubic feet in volume, manufactured by Kinergy Corporation). The G-1 trommel oversize feeds a second 11' diameter, 34' feet long trommel (G-1A) which has plates with 45" inch diameter holes, 7" inches center to center. The G-1A trommel oversize fraction consisting of rags, plastics, cardboard, etc., becomes another RDF fraction. The G-1A trommel undersize is then conveyed to a solid wastes storage bin adjacent to the humus processing section of the plant. This storage bin, designed to hold 11,800 cubic feet of material, is currently being expanded to 18,000 cubic feet capacity.

The stored G-1 trommel undersize is then conveyed to an Organic Removal Jig, (5 feet by 11 feet Remer Jig, by Wemco) which separates the glass and other heavy materials from the light materials such as organics, paper, and plastics. The light materials are dewatered in a Rotating Strainer, (Hycor Corporation) and then fed into a Compression Device (Reitz V-Press). After squeezing the free running water out of the light fraction, the V-Press discharge is conveyed to the humus processing section of the plant.

The glass-rich fraction from the Jig is then conveyed to the Rod Mill Feed Storage Tank (1,700 cubic feet-Kinergy Corporation) and is fed into a Rod Mill (4' diameter x 10' long, with a 3' drum feeder - Koppers, Inc.) which crushes the predominantly glass feedstock into particle size of approximately 200 mesh (850 microns). The discharge from the Rod Mill is screened in two vibrating screens (Kinergy Corporation). The screen oversize is fed back into the Rod Mill for regrinding.

The screen undersize is pumped to a desliming cyclone and classifier (Wemco) which separates fine particles from the feed. The deslimed feed, after mixing with an amine acetate solution, is pumped into two banks of four flotation cells (manufactured by Denver Equipment Company) arranged in a series. The amine acetate solution makes the glass particles hydrophobic and air sparged in the float cells creates a froth which lifts the glass particles to the surface. The froth-floated glass is then conveyed to a vacuum filter (Dorr-Oliver, Inc.) where it is dewatered and then dried in a 4' diameter x 20' long Rotary Dryer (Bethlehem Corporation).

The dryer is fired with No. 2 fuel oil and the glass is discharged at a temperature of 80 degrees centigrade. The dried glass is then fed to a high intensity magnetic separator (Eriez Magnetics) for removing any fine particles of iron. The material settling in the flotation cell (non-glass particles) and the slimes are conveyed to a clarifier where the water is recycled into the grinding and screening operations.

After dewatering in a Rotary Vacuum Drum Filter (Eimco), the sludge removed from the clarifier is more like sand. The sand is disposed in the landfill. The froth-floated dried glass is shipped to a fiberglas manufacturing plant. Gupta[3] describes the process features and costs of this type of glass recovery system. The system as designed and operated yields a low rate of glass recovery. Between 30% to 40% of the glass is drawn off with the RDF fraction. A modification is under study to increase the glass yield.

Nonferrous Recovery

Figure 5 shows the arrangement for nonferrous metals recovery system. The solid wastes heavy fraction (after removal of ferrous and glass fractions) is next conveyed to a Secondary Air Classifier (Iowa Manufacturing Company). This classifier is designed to remove a nonferrous (mostly aluminum) rich fraction which feeds a trommel. The trommel undersize passes through a drum magnet and becomes the feedstock to the nonferrous removal system designed by Raytheon Service Company.

This system consists of stainless steel ramps inclined at 45 degree angle. Permanent magnets of alternating polarity are embedded underneath the stainless steel skin. The nonferrous fraction is then allowed to slide down the ramp. The eddy currents induced in the nonferrous particle make it move at right angles to the direction of the magnetic field, thereby sliding at a 45 degree angle. The nonmetallic particles slide straight down the board. Two successive stages of nonferrous separation yield approximately 50-60% pure aluminum from the feedstock to the ramps. Nonferrous particles which roll are not separated. The escaping nonferrous fraction is directed through a high intensity eddy current separator (Recycal). The yield of nonferrous fraction has improved to 70%. The plant has achieved 90 tons/month peak production during the summer of 1987 (the period of maximum beverage consumption by the public).

Composting System

The humus compost processing section of the plant is shown schematically in Figure 6. At the Wilmington Wastewater Treatment Plant, the primary and waste activated sludges are thickened and digested in four high rate anaerobic digesters. A Belt Filter Press Plant converts the digester discharge into 20% solids filtercake for conveyance to the Delaware Reclamation Project.

The sludge from the treatment plant is stored in three sludge receiving hoppers (Fairfield Company). These hoppers have a capacity for holding 45 tons of sludge. The remaining fraction of the solid waste is then conveyed to a Gundlach Cage Mill where it is blended with sewage sludge from the storage hoppers. The solid waste/sewage sludge ratio is 1 to 1 (wet basis). The homogenized mixture (at an average moisture content of 55%) is fed into four Fairfield Digesters. Each digester has a diameter of 100 feet and an operating depth of 6 feet. These digesters have the capability of composting the solid waste/sewage sludge mixture in an aerobic environment. The digester consists of a rotating bridge, one half fitted with mixing augers which move the solid waste material from the periphery of the circular tank to the center of the tank. The other half of the bridge acts as the feed mechanism for the sewage sludge/solid waste mixture. The tank itself is covered with an aluminum dome to prevent precipitation from affecting the moisture content. The sewage sludge/solid waste feedstock has a residence time of approximately five (5) to seven (7) days in the digesters.

The composted material is removed from a central annular space through conveyors to a coarse screen. The screen undersize is dried in a Heil Triple Pass Air Drier. The drier is initially fired with No. 2 fuel oil. The dried humus is fed to a compost screen which makes an initial separation of non-compostable plastics and other materials from the compost. The screened compost is further refined in another vibrating screen. The screen undersize is recycled to the compost drier system to supplement the fuel for providing the heat necessary for drying the compost.

The dried compost, after screening, is stored in bins for shipping to users. The primary markets for the compost are in the agricultural and horticultural sectors of the state. One primary market was the broiler industry where the compost was expected to be used as poultry litter (an absorbent material for poultry wastes in broiler houses).

In Table 4, chemical composition tests of the compost show the presence of polychlorinated biphenyls (PCB) and chlordane in addition to the usual array of heavy metals. Because of the PCB levels (1 to 6 ppm) the Authority has voluntarily withdrawn DRP compost from the litter market even though a limited growth test showed that birds grown on the litter containing PCB did not significantly absorb the compound in their tissue or fat. The presence of PCB is puzzling because that material has not been manufactured or sold in the U.S. since 1972. Two years of testing has shown that PCB concentrations, apparently attributable to the solid waste, follow a cyclic pattern with low values in winter months and higher values in late summer and fall months. A third year test is expected to confirm this pattern. The states and the U.S.E.P.A. have not adopted uniform criteria and application rates for compost materials. Consequently, the interstate marketing of compost for horticultural uses is also hampered by bureaucracy.

While awaiting the state and federal regulations for distribution of compost into the horticultural and agricultural markets, the Authority is using the annual production of 60,000 tons of compost for its landfill operations as intermediate cover and also for development of slope vegetation (in lieu of conventional topsoil).

Energy Generation

The RDF is shipped to the EGF where it is combusted with unprocessed solid wastes. The plant uses a modular combustion system (VICON, Inc.) with waste heat boilers and electrostatic precipitators. This cogeneration plant supplies approximately 400 million lbs. of steam (@ 230 psi and 400°F) to the ICI Americas Atlas Point Plant. The EGF has two 9 megawatt General Electric turbine generators. The minimum annual power production after satisfying internal needs is 54 million KWH. The power is supplied first to the Authority's DRP (12.6 million KWH/year). The excess is sold to Delmarva Power Company at 95% of the utility's energy avoidance cost (currently 3.1 cents/KWH). The EGF is owned and currently operated by the General Electric Credit Corporation.

The EGF started continuous operation in November, 1987. The preliminary data from this plant indicates that the environmental conditions for air emissions as well as the ash quality can be met by this plant. There is considerable discussion today about ash disposal problems because of the potential for leaching metals and organics from the ash. While starting up the Vicon units, broken wooden pallets and scrap wood were used for building up the furnace temperature. There was an opportunity to obtain ash samples when only wood was being combusted. Table 5 compares metals content of ash from burning wood and the ash from waste combustion. Arsenic, Copper, Chromium, Lead, Mercury, Silver and Zinc concentrations for waste fly ash are higher than those for wood fly ash. This is normally anticipated. However, the concentrations of Cadmium and Nickel for wood fly ash are in the same ranges as waste fly ash! Selenium is one metal which shows a higher concentration for wood fly ash as compared to waste fly ash and the reasons for such an anomaly are not obvious. More sampling and analysis may provide some clues.

Table 6 compares select metals concentrations in the wood fly ash with the proposed Toxic Characteristics Leaching Procedure of the U. S. EPA. Wood fly ash fails the test for Cadmium and Lead. This is not the case for waste fly ash.

Table 7 shows the same comparison for waste mixed ash which passed the test although lead values approach the regulatory limit.

Table 8 provides an indication that materials recovery up front may improve the ash quality. More tests are needed to support this conclusion.

The Delaware experience indicates that it is feasible to recover materials from the solid waste stream. By providing such a central processing plant, the users still have the option of source separation. Since source separation programs are not very successful, those who fail to separate the materials will supply the plant with the waste fraction. Therefore, a central processing plant becomes the secondary net for capturing recyclable materials. The benefits derived from this are improved ash quality, assured recycling of marketable materials and energy savings.

The Delaware experience also indicates that the materials recovery program <u>does</u> have significant operating problems. The DRP has demonstrated that it is possible to <u>resolve</u> these operating problems with persistence, patience and a pragmatic approach - <u>the P^3 Rule of solid waste</u>. In solid waste processing, there is no real control on the quality and quantity of the feedstock, and the plant must be capable of adjusting to the considerable variations. The lessons learned from the Delaware program suggests the following future improvements:

*   Trommeling of the solid waste first and developing systems for breaking the plastic bags will be helpful in recovering more glass, ferrous and non-ferrous fraction.

*   Following such trommeling, additional separation steps to recover light paper and plastics fraction are feasible, particularly after size reduction. Such separation can take the form of air classification or further trommeling. The resulting fuel fraction will be a cleaner fuel and, therefore, produce ash of lower metals content.

LESSONS LEARNED

*   The combustion systems must be specifically designed to use the RDF. The idea of supplying this to utility boilers for co-firing with coal is not very attractive. The successful systems are those where the boilers were properly designed or modified to accept RDF.

*   The ferrous fraction can be very efficiently removed (70% - 90% recovery). The sale of the entire ferrous fraction poses a difficult problem because of the current economic state of the U.S. steel industry. The light ferrous fraction is usable in mini-mills capable of melting scrap. More mini-mills are required if all the light ferrous fraction expected from source separation or other programs to process solid wastes is to be fully utilized. Export market is another possibility, even though it is highly cyclic.

*   Aluminum recovery is becoming very important because of the high price paid for scrap aluminum. The Raytheon ramp eddy current separators in tandem with high intensity electric eddy current separators have shown great promise. Recovery rates of 55% - 70% are feasible with such systems.

Recovery of high purity glass is technically feasible. The cost of production far exceeds the current market value. The sale of glass from such systems is highly dependent upon the nearness of a glass factory. Otherwise, the transportation costs will become prohibitive and glass cannot be marketed. The continued use of glass containers is facing a serious challenge from plastics and aluminum. If this trend continues, glass recovery becomes more difficult.

*   It is possible to co-compost sewage sludge and solid waste on a large scale. The quality of the compost can be affected by the presence of heavy metals and organics such as PCB. Marketing compost requires considerable effort. Quite often it is given away. Interstate regulation of compost creates a bureaucratic problem in obtaining timely permits for such sales. Therefore, composting, while an

essential part of a solid waste disposal system, must rely on local markets for compost disposal. One attractive option is to use the compost on the landfill as a replacement for topsoil or other soil conditioners.

COSTS

The Delaware system is based on a user fee system. No operating subsidies are provided by Federal, State or local governments. Federal and State grants of $46.1 million were provided for the design and construction of the DRP. No grants were provided for the EGF or the landfill.

The user fee for the current fiscal year (July 1, 1987 - June 30, 1988) is $35.26/ton. The same fee is charged at the landfill, the DRP and the EGF. The decision on routing the waste is made by the Authority staff. Table 9 shows the major components of the current user fee.

The user fee can be translated into a monthly cost per family as follows:

* A family dwelling in New Castle County generates 1.6 tons of solid waste/year and pays $56.42/year or $4.70/month.

* The average charge for solid waste collection and disposal service (twice/week) is $186.00/year or $15.50/month.

* Therefore, solid waste collection costs $129.58/year or $10.80/month/family (69.7% of monthly cost).

* Solid waste disposal accounts for 30.3% of the monthly cost.

CONCLUSION

Materials recovery on a large scale through central processing plants is quite feasible. If the goal is to recycle certain materials, then central processing plants offer the most promising method of achieving the society's goals.

Materials recovery prior to combustion of solid wastes offers a means of achieving two goals - recycling and reduction in the metals content of ash. If the new type of landfills equipped with liners and leachate collection systems are used for ash disposal, the perceived threat of groundwater pollution is minimized.

Further improvements in materials recovery are possible and attention should be paid to recovery of certain plastics.

Table 1
CONTENT OF MUNICIPAL SOLID WASTES

		Pounds/Year/Person	Percent
1.	Paper Products		
	a. Packaging Paper, Cardboard	300	25.0
	b. Junk Mail	90	7.5
	c. Newsprint	150	12.5
	Total	540	45.0
2.	Woody Materials (Yard Wastes)	208	17.3
3.	Glass	100	8.3
4.	Food	10	0.8
5.	Ferrous Metals	72	6.0
6.	Plastics	108	9.0
7.	Rubber, Leather & Textiles	15	1.3
8.	All Other	129	10.8
9.	Aluminum	18	1.5
	Total	1,200	100.0

Table 2
EASILY RECYCLABLE FRACTION

	Solid Waste Constituent	Annual Pounds/Capita	Annual Tons/1000 Population
1.	Newsprint	150	75
2.	Glass	100	50
3.	Ferrous Metals	72	36
4.	Aluminum	18	9
	Subtotal	340	170
5.	Yard Wastes	200	100
6.	All Other	660	330
	Total	1,200	600

Table 3
RECYCLABLE FRACTION YIELD

TYPICAL TOWNSHIP
POPULATION: 12,500

	Solid Waste Constituent	Anticipated Yield Tons/Year	Actual Yield Tons/Year	Difference Tons/Year
a.	Newsprint	937.5	500	(437.5)
b.	Glass	625.0	250	(375.0)
c.	Ferrous Metals	450.0	200	(250.0)
d.	Aluminum	112.5	60	(52.5)
	Subtotal	2,125.0	1,010	(1,115.0)
e.	Yard Wastes	1,250.0	1,000	(250.0)
f.	All Other	4,125.0	5,490	+1,365.0
	Total	7,500.0	7,500	-0-

Table 4
SELECTED CHEMICAL CHARACTERISTICS
OF COMPOST FOR 1987

	Parameter	n	Mean	St. Dev.	Range
1.	Cadmium	51	3.4	0.8	1.8-6.3
2.	Chromium	51	203.	38.	132-296
3.	Copper	50	294.	102.	174-634
4.	Lead	51	597.	108.	409-864
5.	Mercury	51	4.9	1.4	3.0-9.1
6.	Nickel	51	293.	102.	167-516
7.	Zinc	51	818.	138.	545-1140
8.	PCB's	51	3.4	1.7	1.0-7.0
9.	Chlordane	51	1.37	0.36	0.7-2.3

NOTE: All values are reported as micrograms per gram - dry weight.

Table 5
ASH CHARACTERISTICS
(ALL VALUES IN MG/KGM - DRY WT.)

	Metal	Wood* Fly Ash	Wood* Bottom Ash	Waste** Fly Ash	Waste** Mixed Ash
1.	Antimony	20.0	14.0	20.0	10.0
2.	Arsenic	72.7	47.1	99.7	22.7
3.	Beryllium	0.5	0.7	0.50	0.7
4.	Cadmium	220.0	175.0	227.0	30.8
5.	Chromium	63.5	74.3	94.2	72.9
6.	Copper	691.0	1,370.0	773.0	609.0
7.	Lead	8,960.0	14,200.0	16,300.0	3,980.0
8.	Mercury	2.8	1.3	7.9	1.60
9.	Nickel	52.2	58.9	53.1	65.6
10.	Selenium	81.9	50.5	9.7	3.6
11.	Silver	16.3	11.8	25.1	10.6
12.	Zinc	8,670.0	6,820.0	9,610.0	2,570.0
13.	Thallium	2.0	3.0	2.0	3.0

*September 1987
**June 1987 - Consists of both bottom and fly ash mixed in the quench tank.

Table 6
COMPARISON OF METALS CONTENT
ASH VALUE vs. TCLP EXTRACT
SEPTEMBER 1987

	Metal	Wood Fly Ash Dry Wt. (mg/kgm)	Wood Fly Ash TCLP** Extract (mg/l)	Proposed Regulatory Level (mg/l)
1.	Arsenic	72.7	0.006	5.0
2.	Barium	–	0.1	100.0
3.	Cadmium	220.0	7.29	1.0
4.	Chromium	63.5	¼0.05	5.0
5.	Lead	8,960	11.3	5.0
6.	Mercury	2.76	¼0.001	0.2
7.	Selenium	81.9	0.071	1.0
8.	Silver	16.3	¼0.01	5.0

**Toxicity Characteristics Leaching Procedure – Proposed by the U. S. Environmental Protection Agency.

Table 7
COMPARISON OF METALS
(ASH VALUE vs. TCLP EXTRACT)
JUNE 1987

	Metal	Mixed Ash Dry Wt. (mg/kgm)	TCLP** Extract (mg/l)	Regulatory Level (mg/l)
1.	Arsenic	22.7	0.004	5.0
2.	Barium	–	2.0	100.0
3.	Cadmium	30.8	0.595	1.0
4.	Chromium	72.9	0.05	5.0
5.	Lead	3,980.0	4.30	5.0
6.	Mercury	1.60	0.002	0.2
7.	Selenium	3.64	0.01	1.0
8.	Silver	10.6	0.01	5.0

**Toxicity Characteristics Leaching Procedure

Table 8
METALS EMISSION OF MUNICIPAL WASTE
COMBUSTION PLANTS IN THE UNITED STATES*
(POUNDS/MILLION TONS OF MSW COMBUSTED)

		Date	Type of Emission Control	Lead	Cadmium	Chromium	Mercury	Arsenic	Beryllium
1.	Hampton, VA	1982	ESP	88,000	4,630	-	20,500	2,160	0.18
2.	Braintree, MA	1978	ESP	85,000	2,620	-	221	253	0.48
3.	Akron, OH	1982	ESP	47,400	1,850	2,440	909	751	-
4.	Niagra, NY	1985	ESP	12,900	530	904	3,160	192	0.96
5.	Albany, NY	1985	ESP	9,460	328	64,700	4,290	186	-
6.	Tulsa, OK	1986	ESP	3,390	-	-	3,580	-	0.03
7.	EGF – DE (April '87)	1987	ESP	2,060	133	561	4,169	20	4.5
8.	EGF – DE (Oct. '87)	1987	ESP	537	31	38	3,674	7	3.9
9.	Marion, OR	1987	SD/FF	292	-	-	2,880	-	0.02
10.	Commerce, CA	1987	SD/FF	33	28	56	3,380	16	5.5

*Data extracted from a report by F. Hasselreiis of Gershman, Brickner and Bratton. This data is available in the U. S. EPA Data Base. ESP - Electrostatic Precipitator; SD/FF - Acid Gas Scrubber/Fabric Filter.

Table 9

		$/Ton	%
1.	Debt Service	9.97	28.27
2.	RDF & Ferrous Recovery	6.81	19.31
3.	EGF Costs	6.20	17.58
4.	Landfill	5.04	14.29
5.	Utilities	1.28	3.62
6.	Insurance	0.95	2.69
7.	Residue Hauling	0.77	2.18
8.	Methane Recovery	0.25	0.71
9.	Environmental Monitoring	0.18	0.51
10.	All Other	2.56	7.26
11.	DSWA Management Fee	1.25	3.54
	Total	35.26	99.96

THE AKRON, OHIO'S RECYCLE ENERGY SYSTEM: AN RDF SUCCESS STORY

David B. Spencer ■ Waste Energy Technology Corporation, Bedford, MA 01730
Linda A. Sowa ■ Department of Public Service, Akron, OH
Robert L. Johnson ■ Waste Energy Technology Corporation of Ohio, Akron, OH

The Akron Recycle Energy System (RES) has completed its second full year under the control of the City and wTe Corporation of Bedford, MA. Major strides were made toward accomplishing the goals set out for 1987—a year in which operational efficiencies have improved dramatically. Refuse-derived fuel (RDF) production increased, the steam customer base was broadened, safe operation became the norm, and the overall financial condition continued to improve. This paper discusses the progress made in the various areas which serve as indicators of the RES's performance. It references the calendar year 1987. Comparisons to 1986 information are made.

BACKGROUND

The Akron Recycle Energy System (RES) is a large municipal solid waste RDF facility with a previously troubled history. For a period of 6 years before wTe took over operation as a public enterprise under the direct control of the City of Akron, the Akron RES was unable to operate successfully for any consecutive 12-month period. It operated substantially below its rated capacity and at a very high net disposal cost. More expensive natural gas was often used to produce steam reliably.

Under joint public control and daily wTe management and operation since late 1985, the Akron RES has enjoyed the longest uninterrupted operating period in its history. Steam deliveries have been continuous. 95% of the steam is produced from firing of refuse compared to 50-60% in past years. The plant is now operating at its rated capacity of about 1000 tons per day. It currently services the City of Akron, 62 private waste haulers, three surrounding communities, Cleveland Heights to the North, and the City of Canton to the South.

The City has received the benefit of dramatic improvement in plant economics and lower waste disposal cost. The facility is now a truly reliable, cost effective, regional waste-to-energy and recycling facility.

System Description

Figure 1 illustrates the overall system design of the RES. Refuse is delivered by packer trucks, dump trucks, and transfer trailers to a receiving area where it is weighed electronically and dumped into a storage pit. The scale system has been modified to integrate it into the plant's fully computerized accounting system. Up to 200 loads per day are received at the RES between 6:00 a.m. and 4:30 p.m. Turnaround time for a vehicle to weigh, tip, and exit the building is about five minutes. Operating personnel inspect each load of waste as it is dumped into the pit to be sure it does not contain unacceptable material.

The facility's major components are: two 1,500 horsepower shredders; a two-stage drum type magnetic separator, miscellaneous conveyers and an ash handling system; three 125,000 pound-per-hour, saturated steam, gas/RDF fired Babcock and Wilcox boilers; and two 2-MW turbine generators. The generator units produce electricity for in-plant use. Exhaust steam from the turbines is used in the deaerator tanks, and to make hot water for plant use. Hot water is also sold to local businesses and a condominium complex adjacent to the plant.

...ly facility in the ...nd co-fires them ...EPA permits to ...stitution ratio of

...n 18-mile steam ... this network, ...30 commercial ...ildings, hotels, ...entral Business ...steam to Akron ...ty, and B.F.

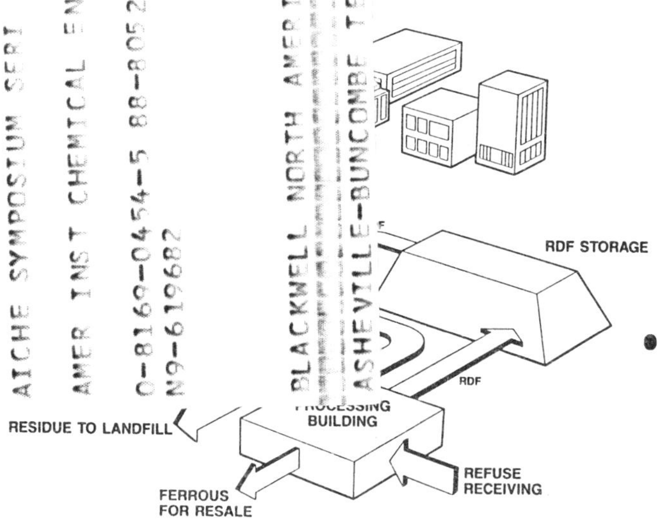

Figure 1. Overall system design—Akron Recycle Energy System.

New projects will produce and sell chilled water, and recycle glass, aluminum, and plastic. The City has now completed purchase of the adjacent B.F. Goodrich steam plant to add to the plant's steam production capability. wTe is now operating that facility for the City and is completing project management of the expansion and chilled water facilities.

1987 Achievements

Noteworthy among the achievements reached in 1987 are the following:

- The automated weigh scale system was completed and placed in operation early in the year. Improvements continue to be made as experience is gained. At year end, new scales and computer equipment installations at the Hardy Road Landfill were being completed, to be followed by software installation, start-up, and training provided by the RES.

- The 1987 waste disposal marketing effort was very successful. The original goal of 800 tons per 5 day week (TPD-5) was substantially exceeded by an annual average of 941 TPD, peaking at 1044 TPD in December.

- The potential for revenue from scrap ferrous was realized through a joint venture between the City and wTe Corporation. About 50 tons per day of ferrous metals produced at the facility are being purchased by wTe under a long term contract. The City received over $17,800 in cash payments from wTe since March, while offsetting an estimated $60,000 in ferrous scrap hauling and landfilling costs.

- The production goal of 200,000 tons, set for all of 1987, was surpassed by almost 25% with a total volume processed in excess of 245,000 tons. Waste received was consistently about 20,000 tons per month.

- Focus on economical procedures for boiler operation was part of the ongoing agenda for 1987. Success in this area can be illustrated by a 75% reduction in the quantity of steam produced from gas since December of 1986 --- yielding an estimated $2.0 million in avoided natural gas costs.

Tremendous strides were made in the effort to expand the RES customer base during 1987. Ohio Edison and B.F. Goodrich signed on as long-term customers, increasing annual steam sales by an estimated $1,000,000 combined. Discussions have begun with the University of Akron concerning the installation of a centralized chilled water plant which will provide increased steam sales. The stability of the Central Business District customer base continued to be aided by the City's downtown redevelopment efforts.

Finally, a variety of procedural and systems automation achievements were realized in 1987. Plant maintenance efforts were bolstered with the creation of a maintenance planner position and equipping the planner with an automated maintenance management data-base system.

Purchasing responsibility shifted substantially to the City. Throughout 1987, the City effected approximately 73% of the RES's monthly purchases. Substantial effort by both RES and City Purchasing Department personnel was dedicated to developing a library of accurate specifications unique to the RES in order to render future purchasing more efficient.

An automated inventory control system, designed to interface with the maintenance management system and RES purchasing was developed and placed in use. Also, work progressed favorably on projects to develop and install automated operating logs, and track system and equipment availability and reliability.

SYSTEM RELIABILITY

In the last 2 years, steam service was lost only once in February, when snow caused a transformer outage resulting in a loss of electrical power. The RES otherwise effectively achieved its goal of uninterrupted service to both steam customers and waste haulers. The system's "interruptable" customers, who pay a lower rate because they have their own back-up boilers, were called to place their units in service only twice when two boiler failures occurred in April 1987. The failures, caused by routine wear on boiler tubes, pointed to the accuracy of the 1987 maintenance plan which called for two boilers to be overhauled during the summer months.

Waste receiving services were interrupted on only two occasions. A shredder explosion in late July that caused minor damage to a roll-up entrance door to the receiving floor made it necessary to turn away a few early arrivals until the door could be opened. In November, two low-volume haulers were diverted to landfills when the tipping floor was filled to capacity.

The City assumed ownership of the B.F. Goodrich steam plant on November 30th. The facility is staffed and operated by wTe Corporation as part of their RES operating activity. With the B.F. Goodrich plant as back-up, the reliability of RES steam service and the ability to service a much broader customer base can be assured.

SAFETY

Safety continues to be the hallmark of the RES. Although waste processing has some inherent risks, through the exercise of control over the materials processed, and by assuring adequate safety facilities and procedures, the outstanding safety record accumulated by the RES since its reopening in 1985 has been sustained. No industrial injuries or explosions serious enough to interrupt production occurred during the year.

RES receiving floor personnel began spot inspections of hauler loads dumped directly on the receiving room floor for hazardous and otherwise "unacceptable" wastes. Several potentially dangerous items were pulled from the inspected loads and three haulers had their tipping privileges temporarily suspended for delivering "unacceptable or unapproved" waste.

A review of all "approved" waste generators was conducted, followed by site visits where appropriate, to re-establish acceptability of waste streams. Coincidentally, an automated database of generator/hauler/waste contents information was created as a customer service tool.

Documented fire and evacuation plans were formulated, approved by the Fire Department, and put into practice, helping to assure the safety of all RES personnel and visitors during emergencies.

Liability insurance was successfully obtained by wTe for the RES earlier in the year ending a period of self-insurance thus protecting the City from excessive exposure to financial losses. Property insurance and business interruption insurance were obtained in 1987, for the first time since the 1984 explosion, demonstrating the success of the investments in safety-related capital improvements.

CAPITAL IMPROVEMENTS

The 1987 RES capital improvement program, budgeted at $250,000, was comprised of six projects. Three projects targeted operational improvements . . .

- The installation of an uninterruptable power supply to guarantee electricity during emergencies was completed and operational in November.

- A process cooling water recirculating system was installed earlier in June.

- Remote equipment monitors, for use in areas where personnel are prohibited during waste processing, were purchased for installation in early 1988.

One responded to the structural maintenance needs of the building . . .

- The installation of steel armor plating over the structural beams supporting the receiving floor was placed under contract in late 1987 for completion during the first quarter of 1988.

Another addressed safety of personnel . . .

- Control room ventilation improvements, to ensure a breathable environment during emergencies, were contracted for construction in early 1988.

And, one reached out to the neighbors of the RES . . .

- The installation of a structure designed to reduce steam venting noise was scheduled for March of 1988.

In addition to the programmed capital projects for 1987, a $2.3 million project was begun which, when complete, will supply steam and chilled water under a 10 year contract to the B.F. Goodrich headquarters and Adhesives Division facilities. The project has been sized and laid-out to accommodate anticipated redevelopment of the remaining B.F. Goodrich property.

AVAILABILITY OF SOLID WASTE

Original estimates of 800 tons per day (TPD-5) were surpassed by a healthy margin, with the year-end average of 941 TPD-5. Significant increases over the 1986 volume of waste being delivered by the Cities of Cuyahoga Falls and Medina, coupled with the continued deliveries from the City of Cleveland Heights and a 60% increase in small-hauler usage, contributed to this year's sustained average intake of 20,470 tons per month. By contrast, 1986's waste intake averaged 15,520 tons per month.

Figure 2 illustrates the amount of waste received during 1986 and 1987. In 1987, 38% of all waste received came from City of Akron residential collection, 23% from other municipal customers, and the remaining 39% from commercial and independent private haulers.

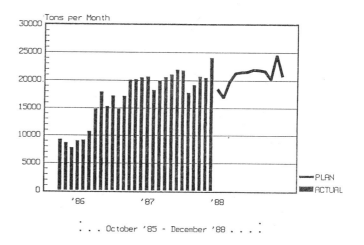

. . . October '85 - December '88 . . .

Figure 2. Waste received.

PRODUCTION

The RES consumes municipal solid waste and tires, processes them into refuse-derived fuel (RDF), and produces electricity for internal plant consumption, hot water for sale to commercial customers, and steam for plant use and for commercial sale where it is used for both heating and cooling. A new capital improvement program to produce and distribute chilled water is underway. A comparison of 1986 and 1987 production levels is given in Table 1.

Table 1. Production summary.

	1986	1987	INCREASE
Waste Processed @ RES - Solid Waste (tons) - Tires (tons)	186,243 183,456 2,780	245,639 242,688 2,951	+32% +32% +6%
Tires Shredded (tons)	5,626	5,554	-1%
Steam Generated (Mlb)	1,754,695	1,769,411	+1%
Steam from RDF (%)	75	94	+25%
Steam Sold (Mlb)	612,000	744,100	+22%
Hot Water Sold (mBTU)	2,998	4,333	+45%
Ferrous Scrap - Landfilled (tons) - Sold (tons)	9,249 0	962 8,585	-90% Infinite

The total volume of waste processed in 1986 was 186,243 tons, averaging 714 tons per day (TPD-5). In 1987, the RES processed a total of 245,639 tons of "acceptable waste" averaging 941 tons per day (TPD-5).

The Hardy Road tire shredding operation processed 5,554 tons of tires as compared to 5,626 tons in 1986. This amounts to over one-half million used tires. Tires are burned at the RES as needed to supplement the RDF in quantities of no more than 10% of the aggregate RDF by weight. The amount of tires burned at the RES in 1987 was slightly less than 3,000 tons.

The amount of steam sold in 1987, 744,100 Mlb, was 22% ahead of the 1986 sales volume of 612,000 Mlb. The re-introduction of B.F. Goodrich as a major customer is expected to add an estimated 264,000 Mlb per year to sales. Adding the 310 Mlb/hour of steam generating capacity of the B.F. Goodrich steam plant to the RES's 390 Mlb/hour provides ample back-up for the immediate future, as well as a solid foundation for expansion of steam sales for years to come.

Boiler utilization on RDF in 1987 was considerably greater than in 1986, averaging 16.1 hours per day versus 12.3 hours respectively. A boiler is not utilized on RDF when there is inadequate steam demand for all three boilers, when there is inadequate waste supply, or for periods of scheduled and unscheduled outage. Availability and reliability were not measured. The increased RDF utilization time of the boilers indicates a higher quality of maintenance, permitting a greater consumption of RDF and a consequential reduction in the cost of gas as an alternative fuel.

REVENUES

As shown in Table 2, revenues for 1987 totaled $7.3 million, compared to 1986 revenues at $6.7 million. Actual revenues in 1987, driven primarily by steam sales, trailed original estimates by approximately 18%, due in part to a warmer than normal weather pattern that similarly affected 1986 steam sales and significant reductions in energy prices based on oil and gas prices. Slumping steam sales accounted for 77% of the 1987 shortfall. Tipping fees were held to $8.25/ton in 1986 and only $10.00/ton in 1987 in order to compete with low cost landfills in the area which charge $11.50/cubic yard. Due mostly to increases in production and not tipping fees, tipping revenues fared better against 1987 projections exceeding 1986's by almost $700,000. Hot water sales finished the year at about 150% above initial estimates and better than twice 1986 actuals.

Table 2. Revenue summary.

REVENUES	1986	1987	INCREASE
Tipping Fees - Solid Waste - Tires	$1,582,000 36,000	$2,191,000 51,000	+38% +42%
Sales - Steam - Hot Water - Ferrous Scrap	4,803,000 21,000 1,000	5,003,000 45,000 18,000	+4% +114% +1800%
Miscellaneous	301,000*	2,000	N/A
TOTAL	$6,744,000	$7,310,000	+8%

* Includes an explosion insurance claim amount of $167,000 plus income of $134,000.

The addition of B.F. Goodrich as a regular and substantial steam customer for the final five months of the year added a total of $336,000 in new revenue. Projected 1988 B.F. Goodrich steam and chilled water sales add over $1,000,000 to the RES's revenue base.

In addition, Ohio Edison, while making a 10 year commitment to purchase the RES's excess steam, added another $72,000 to 1987 sales. The Ohio Edison contract is expected to yield an additional $240,000 in 1988.

Figure 3 illustrates the cumulative net income for the Akron RES.

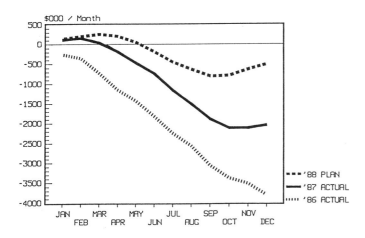

Figure 3. Net income (loss).

COST

RES costs for 1987, at $9.4 million, were well below those of 1986 which totaled $10.6 million. A significant dedication of labor early in the year was needed to hasten the on-line operation of the $2.4 million interim storage facility, which was successfully completed in December of 1986. By year end, that effort, combined with changes in boiler operating procedures, paid off in a reduction in natural gas costs of almost $2.0 million from 1986's actual cost.

Table 3 gives a comparison of costs for the two years.

Table 3. Cost summary.

COSTS	1986	1987	INCREASE
Plant Labor	$3,278,000	$4,007,000	+22%
Subcontracts & Materials	2,735,000	2,769,000	+1%
Utilities	1,320,000	1,540,000	+17%
Natural Gas	2,550,000	548,000	-21%
Project Administration	450,000	242,000	-54%
Management Fee	249,000	247,000	-1%
TOTAL	$10,582,000	$9,353,000	-12%

OVERALL PERFORMANCE

The overall economic performance of the RES in 1987 has been consistent with projections made to the City in late 1985, and again in November of 1986. Table 4 summarizes the project economics. The 1987 net subsidy of $2.0 million, although adversely affected by poor steam sales, is significantly better than 1986's $3.8 million subsidy. A projected subsidy of $500,000 in 1988 is attainable if weather patterns return to normal and the B.F. Goodrich plant operating cost and natural gas projections meet plan. As of mid-1988, these projections were on-target.

Table 4. Project economics.

	1986	1987	INCREASE
Tons Processed	186,243	245,639	+32%
Revenue - Total $ - $ Per Ton	$ 6,744,000 $36.21	$7,310,000 $29.76	+8% -18%
Cost - Total $ - $ Per Ton	$10,582,000* $56.81	$9,353,000** $38.07	-12% -33%
Debt Service	Bonds Defeased	Bonds Defeased	N/A
Net Income (Cost) - Total $ - $ Per Ton	($3,838,000) ($20.60)	($2,043,000) ($8.31)	-47% -60%

* Includes maintenance of 18-mile steam & hot water distribution systems, and tire shredder operation.
+ Includes B.F. Goodrich backup boiler operation.

Economic performance can be expected to improve with increased revenues that are expected to continue to grow as . . .

- Downtown redevelopment continues, adding to the heating and cooling steam markets,

- The RES expands its access to various recyclable materials markets, and

- Area landfills continue to be retired, driving up the cost of refuse disposal for the surrounding communities and thereby increasing tip fees that can be competitively charged to customers.

RES costs have been brought under better control than in previous years and continue to be impacted by major improvements in operational practices. Efficient use of natural gas, the ability to store RDF in the new interim storage facility, better explosion control systems that minimize damage, and more effective manpower utilization techniques have all contributed to 1987's reduced costs. Further improvements are anticipated in 1988 as the project and its managers increase in experience and move down the operating learning curve.

SUMMARY

The RES has continued to operate safely, efficiently, and reliably, meeting both steam production and waste disposal demands of its customer base. The net operating loss of the RES has been significantly reduced from that of 1986. Revenues have been increased and the future revenue base looks strong with the establishment of long-term relationships with B.F. Goodrich and Ohio Edison. Costs have been reduced and are under control. The plant has once again become an insurable risk. Natural gas is no longer a major expense. Markets for recyclable materials have been accessed, adding new revenue and lowering the quantity of non-combustible material landfilled, along with its associated cost. New markets for additional recyclables continue to be sought. Internal systems and procedures are being upgraded to better equip the operations and support staff to improve efficiency.

METRO DADE COUNTY RESOURCES RECOVERY RDF FACILITY: PAST, PRESENT AND FUTURE

Charles R. Strong ■ Montenay Power Corp., P.O. Box 6430, Miami, FL 33152

Past History of the Plant

Metro Dade County's Resources Recovery Facility in West Miami, Florida was permitted for construction and operation in 1979. At that time, the County entered into a long term contract with Parsons and Whittemore. P & W were contracted to build and operate a plant capable of processing 920,000 tons/year, or about 50% of the County's Municipal Solid Waste, into Refuse Derived Fuel. The RDF would be burned to produce superheated steam, drive turbine generators and generate electricity for plant use and for export to Florida Power & Light Company (FP&L).

In 1981 the plant began operation. The front end process had two separate systems to produce RDF and separate non-combustibles. The garbage was received in a building directly into a process pit, fed into a process building to a hydrosposal process system (adapted from the paper industry), and water was then added to the garbage as it was processed in a hydropulper and presses - yielding a 50% moisture RDF. The process also included a mineral recovery system which extracted ferrous, non-ferrous, glass and non-combustibles from the garbage. Material that was classified as trash (tree limbs, cardboard, pallets, tires, etc.) went to an open tipping floor. This material was processed through a Hammermill Size 64 shredder (dry process) and conveyed through magnetic separation, a degritter, and an air knife to produce RDF. Ferrous and non-ferrous materials collected from both processes were sold to scrap dealers. Other non-combustibles and rejects were disposed of in the landfill. The RDF from both processes went into a fuel storage building.

The RDF was then conveyed to four 40-ton/hr. Fives-Cail-Babcock water tube boilers originally designed to burn bark. Their capacity was overstated. Superheated steam was produced in the boilers and drove two 38.5 MW Brown Bovari turbine generators. Electricity that was generated supplied the plant's own needs (40%) and the balance (60%) was exported to FP&L. Chlorinated cooling water for the turbine was supplied by a 6 cell cooling tower. Boiler feedwater was treated wellwater. Combined fly and bottom ash from the boilers was either sold to a cement company or taken to landfill.

P & W operated the plant from 1981 to June 1985. After getting the operation running following a shakedown period, the plant did

process 18,000 tons per week for a short time. However, several major problems became very apparent after the first couple of years:
- Odor from the wet garbage process was a problem for the employees and the surrounding neighborhoods.
- Repeated complaints generated negative publicity for the plant.
- Air pollution from the boilers also brought regulatory complaints.
- Open conveyors allowed RDF to blow all over the plant and yard.
- Contaminated water runoff became a problem.
- The longer the 50% moisture RDF was processed and burned, the more the equipment corrosion and erosion became apparent in the wet process and boiler areas.

Operating efficiency suffered, maintenance costs climbed, and outside complaints continued. In 1984, Dade County officials decided the deteriorating plant operation and reputation were not improving and a change in operators was needed. A termination settlement of the P & W contract was negotiated by the County.

After a solicitation for proposals, Montenay Power Corporation (MPC) was contracted to operate the plant beginning in June, 1985. MPC operated the plant under two 1-year contracts with extensions from June, 1985 to November 1, 1987. During this period, the operational, maintenance and equipment needs were evaluated to upgrade the image and productivity of the plant. Much work was done to improve the wet process and boiler efficiency.

But in September 1986, due to continuing odor, water contamination and maintenance problems, the wet process for garbage was shut down. Garbage and trash were delivered to a common tipping floor and processed through the dry shredders. The workforce was reduced by 150 to 185 people. Productivity was set using the two dry process shredders at 12,000 tons of MSW a week until front end capacity and boiler efficiency could be increased; the same level of production as had been achieved with the wet and dry processes combined. An extensive clean-up of operating and surrounding areas of the plant resulted in a drastic reduction of odor problems. In the last quarter of 1986, the boiler ESPs were rehabilitated and a third field was added to each of the four units. This helped to eliminate the air pollution complaints.

From October 1, 1986 to October 1, 1987, 570,000 tons of MSW was processed through the dry system and the RDF was burned in three inefficient boilers. Nearly 172,000 MWh of electricity, an average of 412 KWh per ton of RDF, were generated with 67% exported for sale to FP&L. The fourth boiler was shut down in January 1987 for extensive repairs.

In 1987, a comprehensive Capital Rehabilitation Program was developed by MPC and the County. Following several months of negotiations, a long-term operating contract was awarded to MPC, effective November 1, 1987. Under this agreement, Montenay will design, finance and implement over $45 million of Capital Project work to return the plant to original permitted capacity with refurbished boilers and a dry front end process. At the same time, all operations will be enclosed and the overall site aesthetics upgraded. The contract calls for a 26-month completion date with rigid deadlines for each of the separate major projects.

Present Plant Operation

With the start of the long-term contract on November 1, 1987, the plant is processing 12,000 tons of MSW through the dry shredders each week. At the same time, the two year Capital Improvement Program has started to restore original capacity and plans to eliminate all regulatory problems by January 1990. Public relations are much improved and looking forward to continued improvements.

Current operation has garbage and trash scaled by the County and deposited on a common open tipping floor. The plant equipment operates 7 days per week - 24 hours per day. MSW is shredded after non-processible elimination. The shredded material goes through magnetic separation where ferrous is separated and conveyed to bins for sale (4% of input). Non-ferrous is not recovered at this time. A degritter and air knife then separates rejects for landfill and the remaining RDF goes to fuel storage. From fuel storage the RDF is conveyed to the three usable boilers.

During the next two years, we will probably average two boilers running full time. This will allow 9,000 tons of RDF to be burned per week, resulting in 3,500 MWh per week of generated electricity. In addition, 2,500 tons per week of combined wet ash will be landfilled. Large metal and white goods will be separated for sale without shredding. Revenues to MPC include a tipping fee for accepted MSW, by-product sales and electricity sales. With the help of the Capital Improvements in progress, the revenue stream will be improved in the future. Cost results for the plant are and should continue to be positive.

In conjunction with the operations, the following major plant improvements are in progress and will be done over the next two years.

1. Complete rehabilitation of the four boilers, including controls, doing one every 6 months over the next two years. Guaranteed RDF burning capacity of 28 tons/hr. per boiler - about a 20% increase over actual productivity of original boiler.
2. Demolition of the old wet process equipment for garbage and installation of two new dry shredding lines with primary and secondary trommels, magnetic separation, aluminum eddy current separation, and reject removal. This will enable a return to separate garbage and trash processing.
3. Refurbishment of the two existing dry shredding lines and trash processing systems.
4. New trash receiving pit and erection of a new trash receiving and processing building will allow all waste to be processed in enclosed structures.
5. Replacement of the boiler's fuel feed system.
6. Installation of new enclosed ash conveyors on all boilers and erection of an ash storage building.
7. Vehicle maintenance moved into an enclosed building from an outside roofed slab.
8. Refurbishment of existing buildings' siding.
9. Odor control system from garbage pit and fuel storage to be ducted for use as boiler makeup air.
10. Establishment of a tire disposal program.
11. Installation of a scrap metal upgrade process. This will include shredding, classifying and/or baling prior to sale.
12. Landscaping and roadway improvements.

In July 1987 the plant was connected to a Dade County sewer line eliminating the on-site use of leachate ponds and constant trucking of liquids.

Future Plant Projections

In 1988 it is projected that 624,000 tons of MSW will be processed. The turbo-generators will generate 185,000 MWh of electricity with 126,000 MWh being sold to FP&L. Over 18,000 tons of scrap ferrous will be sold and 230,000 tons of residue and rejects will go to landfill. By October 1988, all waste and ash will be processed in a fully enclosed system in compliance with all regulatory requirements.

In 1989, it is projected that 728,000 tons of MSW will be processed. The garbage and trash processing will be separated again with the completion of the new garbage shredding lines. By the end of the year, all four boilers should be rebuilt to give optimized combustion efficiency and supply the turbines higher pressure and a higher temperature superheated steam. Electricity generation should be 215,000 MWh with 147,000 MWh sold to FP&L. An estimated 30,000 tons of scrap ferrous, 1,000 tons of aluminum and 52,000 tons of ash will be sold with 190,000 tons of residue and rejects going to landfill. By the end of the year, the various improvements to the plant's appearance will make it a pleasant sight for the public.

In 1990, the original capacity of the plant, 920,000 tons, will be processed (50% of Dade County's MSW) with all new processing equipment in operation. An estimated 364,000 MWh of electricity will be generated and 260,000 MWh will be sold to FP&L. Over 46,000 tons of scrap ferrous, 4,500 tons of aluminum and 100,000 tons of ash will be sold with 130,000 tons of residue and rejects going to landfill.

All of these future projections portray the strong belief that this RDF plant will be an economic, law abiding and aesthetically pleasing facility - a vital part of the Dade County waste disposal program.

COLUMBUS, OHIO'S REFUSE DERIVED FUEL EXPERIENCE

Henry A. Bell ■ Division of Electricity, Department of Public Utilities and Aviation, City of Columbus, Columbus, OH 43215

Today the Columbus Trash Burning Power Plant represents a completely successful operation and serves the city with a reliable method of trash disposal, and will do so for many years to come. As of February 1988 over 1,780,000 tons of trash were used to generate over one billion kilowatt hours of electricity since the plant began operating. Achieving this status today has been a difficult process. The many steps taken toward progress have required innovation in design and hard work and dedication of the employees of the City of Columbus, with support from the Columbus Mayor, The Honorable Dana G. Rinehart.

The Columbus Plant is located approximately four miles south of the center of the city. The immediate area is surrounded with quarrying operations, closed landfills, waste water treatment facilities and a rendering plant. The site is convenient to the Columbus freeway system.

The plant layout covers approximately 52 acres and rests adjacent to a 180 acre lake owned by the American Aggregates Corporation. Through a special arrangement, the lake is used for cooling water for the three 30 megawatt steam turbine generators. Steam is provided by six Babcock and Wilcox

City of Columbus, Columbus, Ohio

boilers which are fueled with either coal or shredded refuse. On the same site is a large shredder station where refuse is dumped onto a concrete floor, then pushed into two separate infeed conveyors which feed two separate vertical hammermills. The refuse is conveyed through a magnetic separator to remove ferrous material and finally conveyed directly to the plant. The refuse conveyor can either carry the refuse directly to a Sprout Waldron live bottom surge bin feeding a boiler, or into a 6000 ton storage pit. Two high speed overhead cranes move the refuse from the pit into the surge bins, each feeding one of the six boilers. Two variable speed augers at the bottom of the surge bin feed trash into two conveyors at whatever boiler fuel rate is required. The trash is then fed into a splitter which feeds into four chutes dropping into four air swept spouts and finally into the boiler combustion area.

Coal, used as an emergency or supplemental fuel for the plant electric output, is conveyed to six bunkers, which feed into a Detroit Stoker feeder and then into the same combustion area of the boiler that trash flows.

The bottom ash falls from a traveling grate into a quench basin, and is then conveyed to an ash storage area. The fly ash is mixed with water to obtain a quality of material that can be conveyed to the same ash storage area. Ash is finally disposed at the Franklin County Landfill.

In addition to the plant's shredder station which has a full acre under roof, there are three satellite shredder stations located in north, west and east Columbus that also process refuse and magnetically remove ferrous material. The City workers who collect the trash, the packer trucks and the motorized equipment maintenance facilities are also located at the outlying stations.

Electricity from the plant is transmitted at 138 KV and 69 KV to electrical substations also operated by the Division of Electricity for distribution to approximately 9,000 customers, and to street lights, water and wastewater treatment plants and major downtown high rise office buildings.

Now let's look back to the beginning history of this new plant which has operated since December 1983, first at half capacity and now successfully at near full capacity, and without any shutdown.

During the conceptual design phases of the Columbus Trash Burning Power Plant, some technological considerations were made regarding the processing, fuel feeding and ash residue handling of the RDF co-fired with coal in a utility boiler.

It was assumed that:

1. The RDF would be a uniform 4"-5" or less shredded material.

2. The Sprout Waldron live bottom bin could successfully handle the refuse and meter the RDF at a controlled variable rate for various boiler load requirements.

3. The mixture of 80% RDF and 20% coal would optimize the combustion process.

4. The boilers would successfully combust the coal/refuse mixture.

5. The bottom ash residue would be quenched and properly conveyed to a belt conveyor using either a Joseph Martin pusher type design or a bottom drag chain type ash handling system.

6. Because of the carbon content of the fly ash, the boiler combustion efficiency would increase with reinjection.

7. Storage of the collected fly ash could result in fires with the potential for explosion, and the fly ash needed to be treated on a continuous basis to avoid such a problem.

8. The trash and ash could be best conveyed by a rubber belt conveyor.

During the first formal full time operation using RDF as the primary boiler fuel, problems began to grow and identify themselves. The vertical hammer mills did not provide a uniform RDF product shredded to 5" and under. Long stringers, rags, pieces of wire and large pieces of iron that passed through the mill and ferrous separator caused plugs in all parts of the fuel feed and ash removal systems.

The boilers suffered major bottom grate problems due to the co-firing of coal and trash. A hot glowing candy-like substance formed across the ash bed, radiating extreme heat into the stoker grate system. This substance became so hot that ferrous material actually became bonded to the grate bars causing interference with the grate air seals. Either the grates would stall, or the grate bars or seals would break away and fail the system. Material such as aluminum melted through the grates, forming masses on the bottom of the grate bars. This would finally stall the grates and the drive system causing major shutdown of the boiler for repair and clean-up. This process was sometimes costing from a few days to several weeks of down time. Also, glass and aluminum melted and plugged the grate tuyers, which caused poor combustion in a steadily degrading manner.

The most serious and costly problem facing successful plant operation was the entire ash handling system. This included the bottom ash quench basins, which were of a design using a pusher and hoe system to convey the ash out of the quench water.

The quench basins battered themselves apart, part by part, causing major shutdown after major shutdown. The system self-classified wire material until the build-up would stall the pusher or rupture its drive parts and system. Not one part of the drive system could endure, over time, the test of moving this type of bottom ash.

The fly ash dustless unloaders failed soon after start-up. After upgrading and beefing up the design, the system still could not convey a consistently treated fly ash material. The dustless unloaders were intended to handle a variable load, but the output consisted of either a runny-sloppy-wet material, or a dry-dusty powder which spewed out all over the entire plant.

The fly ash blackened surfaces all over the plant and the bottom ash spilled from the quench basins to cover the basement floor. This process resulted in massive continuing repairs to the ash system.

Safety inspections showed the presence of lead and cadmium in the fly ash, making it unsafe for personnel to operate in dusty areas of the plant without special respirators. A health and safety program was immediately established which provided health tests consisting of periodic chest X-rays and pulmonary tests and blood lead level tests. Areas of the plant were posted and employees were issued protective gear.

So far, refuse plugs in the fuel feed system, the hot melting bottom ash, and the ash handling system have been discussed. But there are other details that had to be addressed, most regarding the fuel feed system. It was never assumed that this low quality fuel would burn like natural gas, but it does. Because of the uneven nature of the shredded trash, it tends to pack and stay together in clumps. The horizontal feed augers of the Sprout Waldon bins tend to pack the material, worsening the already uneven density of the RDF quality. Because of this, there was an even greater variation in the flow of material. The output of the bin was uneven enough that when large clumps of RDF flowed, the boiler would be driven to a positive pressure and would "puff". It was even dangerous to open the boiler fire door because of this condition.

The puffing action of the boiler occasionally caused flames to reach back through the fuel feed conveyors and set fire to the trash laden conveyor system. Plant employees learned quickly to respond to such fires but found it safer to leave the conveyor doors open rather than allow trash to accumulate.

Even worse, the swings in boiler combustion caused carryover into the mechanical collector and precipitator, causing hopper fires and resulting damage. This carryover was of such nature that it would produce hard, light colored, rounded masses that we named "moon rocks". The rocks consisted mostly of silica.

Plant employees and engineers from R.W. Beck and Associates identified these problems and recommended solutions. Plant employees immediately began working on the oversize RDF flowing from the vertical hammermills by redesigning the hammer configurations and placement. They achieved a much more uniform product from the shredders located at the palnt site. The cost exchange, however, turned out to be approximately a 20 to 30% greater operating and maintenance expense than that of the horizontal hammermills located at the three satellite shredder stations.

Ferrous material was recognized as a serious problem throughout all conveyance systems, including the boiler. Magnets for ferrous removal were originally designed to separate mostly tin cans because of marketability of that type of ferrous. The magnetic separators were beefed up to get the larger ferrous - the "cannonballs". Now the ferrous product is filled with tramp material and is less marketable, but there is a visible improvement in the trash and ash conveyance systems as a result of the change.

Yet in another area, the Sprout Waldron bins would "rag-up" with a combination of wire, long rags and streamers. Plant employees would have to shut down the bins in order to enter and cut away the material from the vertical screws. Also, the vertical screws and horizontal feed augers would compact the refuse at the side walls causing plugs both at the bins and down stream from the bins. Sprout Waldron studied the system, modified the bin wall design, and reduced the compacting of refuse. However, plugs in the exit-bin continued along with failure of the enclosed-type refuse conveying belts. These belts were designed to feed into the two splitters, dividing refuse into the two pairs of air swept spouts feeding the boiler.

The ragging problem was greatly reduced by the attention to the quality of RDF and also by the addition of metal "stabs" along the vertical screw flytes at the point where ragging occurred. A more even flow of refuse was encouraged by the addition of stabs at the end of the bottom bin feed augers. This, however, was not a solution to problems with the enclosed refuse fuel feed belts carrying the shredded trash to the boilers.

Due to air currents present in the enclosed conveyors, light material was blown off the belt. This material and that from spillage along the length of the belt would completely fill the enclosed conveyor. Once the conveyor was packed with the light refuse, the belt would move off-center and would "trip" the conveyor. Furthermore, the splitter at the end of the conveyor would plug because of its limited opening and drop-angle. This splitter was initially modified by plant employees which resulted in reduced plugging; however, the final solution to the fuel feed problem was ultimately the complete replacement of the enclosed belt-type conveyor system with an enclosed vibrating-type feed conveyor. This was a totally new and innovative approach to the problem. The RDF now moves consistently, reliably and at a much more even rate as a result of the "spreading-out" effect of the refuse being conveyed by vibration. Problems with boiler puffing and carryover into the ash hoppers are also greatly reduced.

Meanwhile, during all of these problems, plant employees invented a rear air seal tuyere which is successfully patented by the City of Columbus. A new grate bar retainer was designed and installed to further reduce loss of grate weights and resulting down time.

During the first two years of boiler operation, the ash reinjection system reduced the boiler sidewall water tube thickness by nearly 50% due to erosion. A measure of carbon content of the reinjected ash indicated little or no carbon having any fuel value. The reinjection system was then removed.

The problem relating to the damage caused by the radiant ash substance was discovered to have been from using a fuel mixture containing the design value of 20% coal. This problem has been resolved by fueling the boiler either with 100% RDF or 100% coal, and no other mixture of the fuels. Boiler operation is now very reliable with much more even output of steam flow. Grate life has more than doubled and better results are predicted.

A major accomplishment was the complete replacement of the ash handling systems for all six boilers while the plant was "on line" and burning record amounts of trash. The retrofit design allowed replacement of sections of the ash system on two boilers at a time, allowing the rest of the boilers to continue burning increased quantities of RDF. The bottom ash systems were changed to a hefty drag chain conveyor design, dropping the ash onto an enlarged rubber conveyor belt. The fly ash system was completely changed using redundant dry drag conveyors discharging into a small silo which in turn discharges at a controlled rate. This allows an even flow of conditioned fly ash to discharge into the ash reclaim bins. The new system works very reliably.

But that is not all. The last and most recent of the serious problems deals with heavy metals, particularly lead and cadmium, which are concentrated in the ash stream. To identify these metals, they are leached using the EP-Toxicity Extraction Procedure to test for Hazardous Waste. The test results vary above the 5 ppm limit for lead and the 1 ppm limit for cadmium set by EPA. By EPA standard this renders the ash hazardous. Through the cooperation of the County Government (operator of the landfill used to dispose of the ash), the City, and the Ohio EPA, limestone sand is mixed at a ratio of three parts ash to one part sand (3/1) with excellent results in stabilizing the quality of the ash. Meanwhile, new methods of resolving this problem are being studied.

The result of our work in Columbus is remarkable. People all over the world have been attracted to tour the trash burning facility. Even with this success, there are many more challenges to come.

One such challenge is the high cost of tire disposal, now more than $3.50 per tire at the landfill. The health problems Statewide caused by mosquito infestation from "tire farms" is serious, yet the fuel value of tires is very high compared to trash and even higher than that of coal. We are cooperating with the Ohio EPA by experimenting with the use of a 5 to 10% mixture of shredded tires with the RDF. The first tests have indicated very positive results. A future program is also being pursued.

Resource recovery is picking up national popularity, especially in Columbus. Since iron, aluminum and glass have no fuel value and cause wear and tear to our plant, there is the need for front end separation of these materials. Various programs are being studied including curbside collection of separated material.

The hazardous ash chemistry problem needs to be solved. It represents a national problem for all incineration systems disposing of solid waste. Use of monofills and combining the ash with other waste products are methods being reviewed. Other methods such as making bricks out of the ash are expensive, but are also being reviewed. There is much to do in solving this problem.

In Columbus, waste disposal (including infectious waste) is seen as a regional problem and discussion of regionalization of the operation of the plant, the landfill, and all other waste disposal programs is underway.

We have come a long way in Columbus, but there is still the need to address the future waste problems of the region.

REFUSE-DERIVED FUEL VERSUS MASS-BURN TECHNOLOGY: OGDEN MARTIN'S HAVERHILL/LAWRENCE, MASSACHUSETTS PROJECTS

David L. Sokol ■ Ogden Martin Systems, Inc., 40 Lane Road, Fairfield, NJ 07007-2615

This paper discusses the technological differences between refuse-derived fuel and mass-burn based on Ogden Martin's experience in retrofitting and operating the Haverhill/Lawrence, Massachusetts RDF project and in designing, building and operating numerous facilities using the Martin Mass-Burn Technology. Each phase of the RDF fuel preparation system and the retrofitting of the boilers is described in detail. When compared on an economic basis, it is less costly to design, construct and operate a mass-burn plant. The latter technology requires less labor, ensures greater reliability, and overall operation and maintenance lifecycle costs are less expensive.

Over the past ten years, as the waste-to-energy industry in the United States has evolved from a new entity to a fairly mature one, there has been considerable debate regarding the efficiency and associated costs of constructing and operating mass-burn and refuse-derived fuel facilities. With many cities and counties nationwide faced with the overwhelming task of finding disposal capacity, the questions loom as to which of the technologies -- mass-burn or refuse-derived fuel (RDF) -- is the most efficient and worry-free.

While one can safely say that both technologies require a certain amount of ongoing maintenance, there are advantages and disadvantages that communities must weigh carefully before entering into long-term construction contracts and service agreements. Some of the many questions include: what are the differences in design and construction costs; which system meets or exceeds its energy guarantees over the long-run; which system best ensures worker safety; and which system is likely to perform most efficiently over the 20-25 year life of the service agreement?

REPAIRING & RETROFITTING AN RDF FACILITY: THE HAVERHILL/LAWRENCE EXPERIENCE

Background of the Haverhill/Lawrence RDF Project

As a full-service systems contractor, Ogden Martin Systems, Inc. provides services in designing, financing, building and operating resource recovery facilities for either public or private ownership.

Ogden Martin offers the Martin Mass-Burn System which includes the Martin Stoker Grate and Ash Discharger. The Martin technology is used in more than 130 facilities worldwide, making it the most frequently employed combustion system in modern resource recovery plants.

In mid-1986, Ogden Martin was contacted by Refuse Fuels, Inc. (RFI), the original developers of the Haverhill, Massachusetts Refuse-Derived Fuel (RDF) plant and its dedicated boiler in Lawrence, concerning the possible sale and takeover of the project. The original facility was financed in mid-1982. Construction commenced immediately, and commercial operation began in late 1984. During the initial two years of operation, the facility encountered numerous operating and financial difficulties. Technical problems included malfunctions and breakdowns of the front-end processing equipment used to make the shredded fuel; low operating availability due to boiler corrosion, erosion and fouling; and major emission problems that caused the facility to exceed its permissable limits. These technical failures kept the facility from meeting its economic goals.

The financial transaction to assume ownership of the existing RDF project and build a new mass-burn facility represented the most complex task ever undertaken in the waste-to-energy industry. The package totalled $263 million comprised of $223 million in long-term debt, issued in four series (two taxable and two tax exempt) through the Massachusetts Industrial Finance Authority (MIFA), and equity contributions totalling $40 million from Ogden Corporation. Prior debt on the RDF facility was defeased through $87.5 million in taxable, fixed-rate bonds, and an $18 million equity contribution from Ogden Corporation was used to repair and retrofit the project.

The new 1,650 tons-per-day (TPD) mass-burn facility was financed through an equity contribution of approximately $22 million from Ogden Corporation and a series of bonds totalling $135.5 million: $31 million in taxable, variable rate bonds; $66 million in tax-exempt, fixed rate bonds; and $38.5 million in tax-exempt, variable rate bonds. The design and construction cost of the new facility is $120 million.

Technical Description

The Haverhill/Lawrence Refuse-Derived Fuel (RDF) System is designed to handle 1,300 tons-per-day (TPD) of refuse. The front-end, RDF preparation system, as designed by the Heil Company, is located in Haverhill adjacent to the landfill site, which is also a part of the project (See Figure 1). The combustion facility and the electric power generation plant are located in Lawrence, next to the Malden Mills Complex, which uses some of the generated steam and electricity.

The RDF preparation section comprises the following areas: refuse receiving area, in-feed conveyors, 1,000 horsepower shredders, transfer conveyors, two-stage magnetic separation, 60-feet long trommels with two- and four-inch diameter holes, various product conveyors (i.e. RDF, undersized and oversized rejects and ferrous materials), numerous transfer and loadout chutes, the RDF storage area and a dust control system. It is a very complicated system and, as is customary in these plants, there is redundancy of all equipment to improve operating availability. To increase the yield of RDF, secondary shredders have been added to handle the oversized rejects, which are mostly comprised of combustible materials.

The processed RDF is then hauled to the Lawrence facility, which also has an RDF storage area, horizontal and inclined feed conveyors, distribution augers, return conveyors, various transfer chutes and a dust control system (See Figure 2). As in the preparation plant, all RDF handling equipment was virtually duplicated for redundancy in order to attempt to ensure a minimum operating availability of 85 percent. Actual availability of the total RDF system is approximately 60 percent. RDF is fed into a dedicated boiler, which was designed by Babcock and Wilcox, via hydraulic rams, inclined feeders and air-swept chutes.

The boiler is a conventional Sterling two-drum unit with an integrated spreader stoker combustion system designed to burn 950 tons-per-day (TPD) of 4,800 Btu/lb RDF to produce 250,000 lb. per hour of superheated steam at 750 degrees Fahrenheit and 650 psig. A four-field electrostatic precipitator (ESP) is used for particulate control.

The RDF boiler (See Figure 3) is located next to the existing steam and power generating plant, which includes three older oil and gas-fired boilers that are being used for back-up. In addition, it also includes five small steam turbines, including back pressure and condensing units. The support systems include the water treatment system, compressors, feedwater heating, deaerators, etc., some of which have been retrofitted and replaced depending upon their condition.

PHASE I:

Initial Operation (1984-1985)

After the plant started up in 1984, it was soon discovered that the RDF boiler system, as designed, could not meet its performance requirements without excessive downtime. Major operation problems included the following:
- Bad combustion conditions which caused the flame to extend all the way into the superheater and boiler bank zone.
- High superheater steam and tube metal temperatures.
- Furnace tube-wall corrosion.

In trying to solve these various operational problems, the boiler vendor undertook the following corrective measures:
- Modification of the overfire air (OFA) system, using three-inch diameter nozzles, comprised of 17 in the front wall and 16 in the rear wall, in addition to the original one-inch diameter nozzles.
- Installation of furnace exit platens/screens (tubular heat exchangers forming a screen), arranged at 18 inches lateral spacing, in an attempt to control the ex-

tremely high flue gas temperatures at the superheater inlet.
- Reduction of the superheater heating surface by approximately 22 percent.
- Installation of Alloy 625 welded overlay on the furnace tube walls, up to a level of approximately seven feet above the lower sidewall headers.

Although these corrections and modifications were partially successful, they did not completely solve the numerous problems experienced with the boiler.

PHASE II:

Operating Experience From 1985-Early 1987, After Boiler Vendor's Retrofits

During the next operating period, the condition of the unit started deteriorating rapidly with the following major problems encountered:
- Tube wastage and failures of the waterwalls above the newly installed Alloy 625 overlay, the new furnace exit platens/screens, the superheaters and boiler bank.
- Chain grate stoker problems, including wear, warping, uneven increase in the distribution air hole sizes, blocking of air holes and grate sections with aluminum and various stoker air sealing problems.
- Fuel maldistribution (furnace left to right) affected by the feeding system, where the incoming inclined conveyors feed the unit from the right side, resulting in biased fuel distribution with light fractions fed to the left side and heavy to the right side of the furnace.
- Decrease of the RDF boiler's operating availability.
- Inability to operate at the design excess air of 42 percent.
- Heavy unburned particulate carry-over from the furnace into the convection zone due to poor combustion, undergrate air distribution and localized high-velocity zones in the furnace.
- Heavy fouling of all the convection heating surface.
- Dioxin and furan emissions, which were several orders of magnitude above the limits permitted by the Massachusetts Department of Environmental Quality Engineering.

During phase two of retrofitting, Ogden Martin and RFI performed numerous combustion and performance tests at 100, 90, 75 and 60 percent load to evaluate boiler performance at various loads. Additional steps included a furnace temperature survey, video filming of furnace conditions to monitor fuel distribution, observation of slagging patterns and flame development, including high velocity and particulate carry-over zones.

The test results, together with operator observations and experiences and several metallurgical reports, were evaluated to establish potential causes for these operating problems. More importantly, all of this data was used to determine necessary corrective measures to be taken.

The most likely causes of the tube wastage and failures were attributed to the following: reducing furnace atmosphere conditions, poor fuel and air distribution, insufficient furnace residence time, high flue gas temperatures and velocities (recorded measurements were as high as 2,000 degrees Fahrenheit at the furnace exit), corrosion caused by fuel constituents and ash mainly attributed to chloride, zinc and lead and high particulate carry-over and insufficient clearances between the sootblowers and the tube banks.

PHASE III:

Corrective Measures To Improve Operating Ability, 1987-88

The pressure of cumulative tube failures forced two major boiler shutdowns, during which several hundred furnace exit platen/screens, superheater and boiler-bank tubes were replaced. Additional steps that were also taken to improve boiler operating availability included:
- Increasing the amount of Alloy 625 welded overlay by approximately 20 feet from the original seven to a total height of 27 feet above the lower sidewall headers.
- Replacing and modifying the superheater lower return bends.
- Extensively shielding sections of the furnace exit platens/screens, superheaters and boiler banks and all tubes in the vicinity of the sootblowers.
- Changing the tube size in several rows of the boiler bank in an attempt to control flue gas velocities.
- Installing new thermal drain valves for the sootblowing system.
- Rebuilding the stoker, replacing all bars, front and back seals, grate support rails, fuel spouts, etc.
- Limiting the boiler load to 75 percent.

To establish potential improvements in combustion performance, which was also needed for reduction of dioxin and furan emissions, it was decided to proceed with a furnace model study. A three-dimensional plexiglass model was constructed at one-sixth scale to conduct isothermal flow tests.

Tests performed using this model indicated that the overfire air system, as modified by the boiler vendor, was still performing poorly. Results indicated that the one-inch overfire air nozzles were ineffective and the three-inch nozzles provided a "blanket" effect with mid-furnace clash and resulting poor penetration and velocity distribution. In addition, the RDF fuel injected into the lower furnace had a tendency to pile in localized areas.

Based on the results of the flow model study, it was decided to completely change the overfire system including installing new nozzles of different sizes, numbers and location and adding a separate overfire air fan. In addition, the fuel spouts were also modified. Following all modifications, the unit was officially tested in September 1987 for particulate emissions, dioxins, furans and nitrogen oxides. These tests confirmed that the unit was in conformance with State of Massachusetts dioxin and furan guidelines. Currently, the RDF boiler operating availability exceeds 85 percent, however, only at a 75 percent load.

THE HAVERHILL RESOURCE RECOVERY FACILITY USING MASS-BURN TECHNOLOGY

The two-unit, mass-burn facility is unique from a design standpoint and will provide Haverhill and the surrounding communities with a safe, reliable disposable solution for many years. Each 825 TPD unit will burn waste in a large conservatively-designed, multi-pass steam boiler providing high operational availability.

To conserve valuable water resources and space, a roof-mounted air cooled condenser was selected to condense the turbine exhaust steam. Steam will be produced at a pressure of 865 psig and a temperature of 830 degrees Fahrenheit. The turbine has a name plate capacity of 46 megawatts (MW). A net output of approximately 41 MW will be delivered to New England Power Company at an energy rate of eight cents per kilowatt hour. The facility will begin start up in early 1989.

In addition, an integrated recycling effort is in place that currently focuses on the removal of corrugated paper from the waste stream. The feasibility of converting the Haverhill RDF facility into a large-scale materials recovery site where paper, glass and aluminum would be separated and sold for reuse is currently being evaluated.

COMPARISON OF MASS-BURN & RDF TECHNOLOGIES

Following are some of the major differences experienced during actual operation by Ogden Martin of these two technologies:
- RDF fuel preparation is much more complex and has, therefore, higher in-house power consumption.
- Because of the larger volume and complexity of buildings receiving and handling refuse, it is more difficult to control RDF plant odors.
- Even if thorough sorting of materials is performed beforehand, RDF front-end preparation of fuel is still more susceptible to explosions.
- RDF plants have higher operating and maintenance costs due to the need for more operating personnel (approximately double a mass-burn plant staff) and higher maintenance expenses associated with shredding and handling systems.
- Wet refuse causes plugging of the trommels and handling systems.
- RDF systems generate less steam and power per ton of refuse delivered since some of the combustible material is lost in the fuel-processing phase.
- To achieve higher availability, RDF processing lines must be redundant thereby increasing the capital costs and O&M expenses associated with installing and maintaining such equipment.
- Combustion air distribution in an RDF installation relies on the use of grate bars with relatively small holes and on the resistance of the fuel bed itself which is not always uniform. On the other hand, mass-burn systems use a grate with separate air-zones so that the air supply to these zones is individually controlled and air distribution within the zone is dependent on pressure drop through the grate.
- Because of the superior undergrate air control, the mass-burn grate is less susceptible to plugging caused by metals with a low melting point, particularly aluminum.
- Since RDF is a more homogeneous fuel, these boilers control steam pressure slightly better than mass-burn units.
- RDF front-end preparation is better suited for reclamation of metals.

In addition to these technical differences, the additional staffing and maintenance costs for an RDF facility directly affect the bottom line.

Operational and maintenance costs of an RDF plant average $50-60 per ton of processing capacity. In sharp contrast, the O&M costs for a mass-burn facility would be in the range of $20-30 per ton of processing capacity. If one were to assume that the RDF and mass-burn plants were located side by side, with equal disposal capacity, the RDF plant would require an additional 50 to 60 employees. As a result, the operational and maintenance costs over the 20-25 year life of the project would double that of a mass-burn plant. In addition, because the equipment for the two RDF sub-systems (waste processing and energy production) is so different, there isn't much opportunity for cross-utilization between employees in the two areas.

Based upon Ogden Martin's experience over the past five years, a newly-designed RDF facility would cost 10-20 percent more to build than a new mass-burn plant. For a 1,000 ton-per-day mass-burn project, with scrubbers and baghouses, the capital costs amount to approximately $90-100 million in 1988 dollars. An RDF facility with the same pollution control system would cost a municipality $100-$120 million.

It seems clear from the data presented that, of the two technologies, mass-burn provides the solution that is more economical and reliable in the long run for disposing of solid waste. In Haverhill, the failure of the RDF plant's performance is the primary reason that the mass-burn plant is being built. As a technology, RDF could not be relied upon. By contrast, the design and operating technology of a Martin mass-burn plant serves to successfully minimize or eliminate problems and in turn, significantly reduces the potential for incurring unexpected costs.

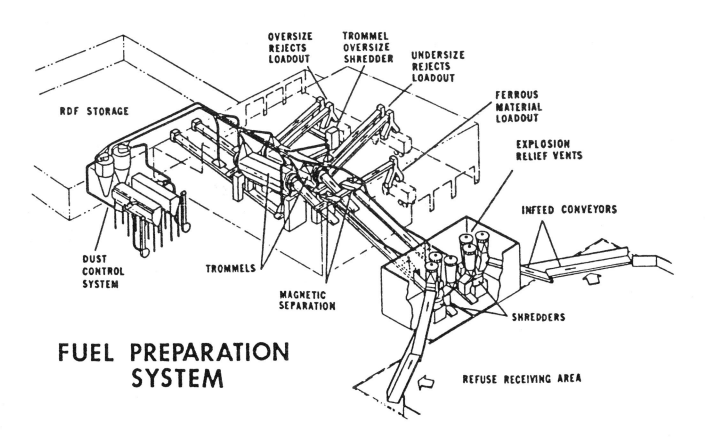

Figure 1. Drawing used by permission of The Heil Company.

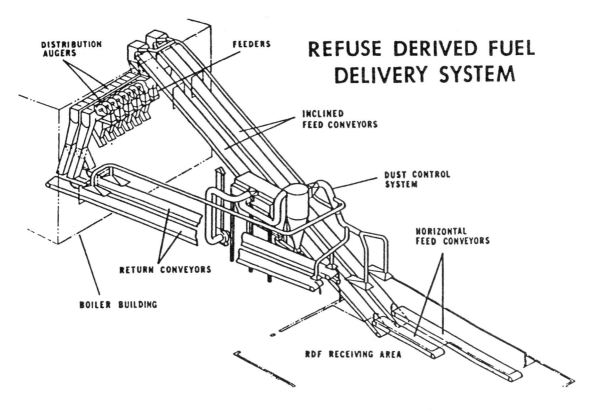

Figure 2. Drawing used by permission of The Heil Company.

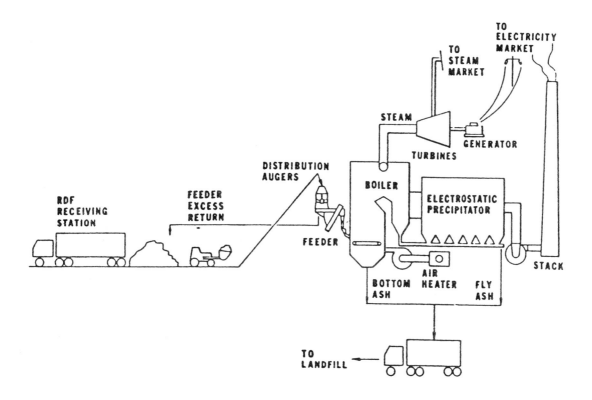

Figure 3. Drawing used by permission of The Heil Company.

RESULTS OF EMISSIONS AND ASH TESTING AT A MODERN REFUSE-DERIVED FUEL PLANT

Francis A. Ferraro ■ KTI Energy, Inc., Four City Center, Portland, ME 04101

In 1987, the Maine Energy Recovery Company (MERC) waste-to-energy facility in Biddeford, Maine became the first refuse-derived fuel plant in the U.S. to operate with a dry scrubber and baghouse air quality control system. During 1987, MERC underwent emissions testing, including participation with U.S. EPA in its emissions characterization program for municipal waste combustors. In addition, MERC has conducted one of the most exhaustive ash characterization programs for a scrubber-equipped municipal waste combustion facility. This paper describes the plant and summarizes the results of these studies.

INTRODUCTION

As in many other states, Maine had identified a number of municipal and private landfills as contributors to groundwater contamination and issued orders that those landfills be closed. As landfill space for disposal of municipal solid waste (MSW) and non-hazardous industrial waste began to became scarce, communities began to look for alternatives to landfilling their waste. The towns of Biddeford and Saco issued a Request for Proposals to construct a waste-to-energy facility for the waste from those cities and a number of other southern Maine towns. In 1983, the contract for this project was awarded to KTI Energy, Inc., and work began on the Maine Energy Recovery Company (MERC) project.

The MERC facility is a 700 ton/day refuse-derived fuel (RDF) waste-to-energy plant which now serves over 30 communities in southern Maine and portions of New Hampshire. The operation of MERC and its sister plant in Orrington, ME (PERC) will precipitate the closing of 7 of the 10 worst municipal landfills in Maine. In all, MERC and PERC will enable the closing of up to 100 landfills.

The purpose of this paper is to describe the MERC facility and discuss the results of the emission tests conducted at MERC and the results of what may be the most exhaustive sampling and analysis program for ash from a modern waste-to-energy facility in this country.

THE MERC FACILITY

Construction on the MERC plant began in September 1984 and start-up activities began in March 1987. The incoming MSW is processed into RDF by a series of shredders and screens. Magnetic separators remove ferrous metals from the shredded waste prior to screening. The RDF is conveyed to either of two (2) Babcock & Wilcox boilers, each with a fuel heat release rate of 150 MBTU/hr. The steam from the boilers powers a single 22 MWe turbine/generator, with the electricity from the plant being sold to Central Maine Power Company (see Figure 1).

Initially, the MERC plant is operating as a multi-fuel plant co-firing wood chips (approximately 20% heat input) with RDF. This co-firing capability allows the plant to operate at maximum power generation, while providing installed capacity for future growth of MSW from the communities under contract to MERC. As the amount of MSW from the communities grows, the amount of wood chips fired will be reduced until the facility fires 100% RDF. Natural gas is also used for start-up and as a supplemental fuel. MERC processes MSW 16 hours/day, 6 days/week, reducing the delivered MSW to approximately 75% of its original weight with a heat content of 5500-6000 BTU/lb, and a size of 90% less than 4 inches (see Table 1).

TABLE 1.
MAINE ENERGY RECOVERY COMPANY
TYPICAL RDF FUEL ANALYSIS
(As Received Basis)

Ultimate Analysis	Percent
Moisture	29.2
Carbon	32.2
Oxygen	24.2
Hydrogen	4.2
Nitrogen	0.4
Chlorine	0.1
Sulfur	0.2
Ash	9.5

Proximate Analysis	Percent
Moisture	29.2
Volatile Matter	52.3
Fixed Carbon	9.0
Ash	9.5
Heating Value (BTU/lb)	5800

For air quality control, the facility incorporates lime-slurry, spray dryer absorbers (scrubbers) and fabric filter baghouses, one train for each boiler. Additionally, the tipping room and processing building are under negative pressure with the air from these areas exhausted through a separate fabric filter baghouse. The filtered air from this baghouse is vented to the inlet of the forced draft fan of the boilers, thereby further minimizing particulate emissions from the facility.

The flue gasses, at about 350°F, enter the scrubbers where they react with a lime slurry (15% wt solids) at about 6-8 gpm. The treated flue gas enters a pulse-jet baghouse which has 6 compartments with 126 fiberglas bags per compartment and a net air-to-cloth ratio of about 4:1. The cleaned gases then leave the baghouse at approximately 280 °F.

The ash from the mechanical dust collector and fly ash and scrubber product from the baghouse are all conveyed to the ash conditioning screws. The wetted ash is then mixed with the boiler bottom ash on the bottom ash de-watering inclined conveyor before being deposited in a trailer for transport to disposal.

MERC EMISSION TESTS

The MERC Air Emission License Application was the first such for a modern resource recovery plant that the Maine Department of Environmental Protection (DEP) had received, and consequently, DEP developed emission limits for MERC based upon the guidelines and experiences of other states. Maine also has a heritage of environmental consciousness, and, as such, the Air

License issued to MERC contained some of the most stringent standards then applied to any resource recovery plant. For example, the emission limits for acid gases were set at 30 ppm with no credit for the efficiency of removal. Also, unlike states that have a Total Suspended Particulate (TSP) limit of 0.01 grains/dscf, DEP required that MERC achieve TSP results that were less than 0.010 gr/dscf.

When the Air Emission License was first issued in 1984 for construction of the MERC plant, it did not contain emission limits nor testing requirements for dioxins. Subsequently, due to public concern over emissions of these trace organics from MSW plants, MERC recommended to DEP that testing for dioxins and furans be included in the Air License renewal which was issued in September 1987. All of the emissions tests were conducted at full load while firing 100% RDF.

While the Protocol for the dioxin test was being developed, it was decided to proceed with the stack emission testing for criteria pollutants. This testing took place during the week of September 23, 1987, and the results are shown in Table 2. Maine also established a permit limit on particulates below 2 microns, therefore, the analyses of the MERC test data included a particle size distribution (see Table 3.)

KTI and DEP both wished to complete the dioxin testing as soon as possible since testing on the MERC stack at an elevation of about 100 feet above the ground during the middle of the Maine winter would be a cold and even dangerous situation, since winds off the ocean could drop the wind chill well below zero.

TABLE 2.
MAINE ENERGY RECOVERY COMPANY
AIR EMISSION LICENSE LIMITATIONS
AND EMISSION TEST RESULTS

POLLUTANT	EMISSION LIMIT [a]	TEST RESULTS [a,b]
NO_x	295	202
CO	400	83
SO_2	30	5
HCl	30	10
VOC	70	0.4
TSP [c]	0.010	0.008
LEAD (lb/MBtu)	0.016	5.5×10^{-5}
MERCURY (grams/24 hr)	3200	105

a) Concentrations in ppmdv adjusted to 12% CO_2

b) Average of four (4) tests

c) Grains/dscf adjusted to 12% CO_2

Additionally, the testing for dioxins and furans at the MERC plant marked the first time that the Maine DEP had been involved in such a test, and because of public concern over emissions of these pollutants from waste-to-energy facilities, DEP asked the U.S. EPA at Research Triangle Park and NY Department of Environmental Conservation to review the testing protocol submitted by MERC's testing contractor for the MERC dioxin test. A local environmental group, the Maine Peoples Alliance, also had the protocol reviewed by their own consultant.

TABLE 3. MAINE ENERGY RECOVERY COMPANY
AIR EMISSION TEST RESULTS
PARTICLE SIZE DISTRIBUTION

	CUMULATIVE PERCENTAGE OF PARTICLES SMALLER THAN THE STATED SIZE (%) (AVERAGE OF THREE TESTS)
0.63	3.4
1.0	6.1
1.25	7.7
2.0	12.9
2.5	16.1
6.0	52.8
10.0	77.7
15.0	91.7
20.0	100.0

The protocol was accepted by DEP with only two minor changes and the test began on December 7, 1987. After some false starts, the test program was completed on December 13, and results were received by the end of January 1988. Based upon the average of the three tests, total PCDD and PCDF emissions were 2.78 ng/Nm3, corrected to 12% CO2 (see Table 4.) The corresponding 2,3,7,8, TCDD toxic equivalents using both the EPA and Eadon toxic equivalency factors were 0.040 and 0.071 ng/Nm3 corrected to 12% CO2, respectively (see Table 5.)

U.S. EPA EMISSION TESTS

As part of its regulatory program to characterize emissions from MSW plants, U.S. EPA's Office

TABLE 4. MAINE ENERGY RECOVERY COMPANY
AIR EMISSION LICENSE LIMITATIONS
AND EMISSION TEST RESULTS
FOR DIOXINS AND FURANS

PCDD	ng/Nm^3 Corrected to 12% CO2 (Average of Three Tests)
2,3,7,8-TETRA CDD	0.0177
TOTAL TETRA CDD (a)	0.0461
1,2,3,7,8-PENTA CDD	0.0151
TOTAL PENTA CDD (a)	0.1425
1,2,3,4,7,8-HEXA CDD	0.0279
1,2,3,6,7,8-HEXA CDD	0.0226
1,2,3,7,8,9-HEXA CDD	0.0176
TOTAL HEXA CDD (a)	0.0868
1,2,3,4,6,7,8-HEPTA CDD	0.1919
TOTAL HEPTA CDD (a)	0.3200
OCTA CDD (a)	0.6226
TOTAL PCDD	1.2180

ALLOWABLE PCDD EMISSION RATE ... 215

PCDF	
2,3,7,8-TETRA CDF	0.0532
TOTAL TETRA CDF (a)	0.6138
1,2,3,7,8-PENTA CDF	0.0426
2,3,4,7,8-PENTA CDF	0.0771
TOTAL PENTA CDF (a)	0.3977
1,2,3,4,7,8-HEXA CDF	0.0624
1,2,3,6,7,8-HEXA CDF	0.0228
2,3,4,6,7,8-HEXA CDF	0.0499
1,2,3,7,8,9-HEXA CDF	0.0093
TOTAL HEXA CDD (a)	0.2758
1,2,3,4,6,7,8-HEPTA CDF	0.1403
1,2,3,4,7,8,9-HEPTA CDF	0.0211
TOTAL HEPTA CDF (a)	0.1693
OCTA CDF (a)	0.1062
TOTAL PCDF	1.5628

ALLOWABLE PCDF EMISSION RATE 65

(a) Totals include the 2,3,7,8-substituted amounts
(b) Mono- thru Tri-CDD and CDF not reported

TABLE 5.
MAINE ENERGY RECOVERY COMPANY
2,3,7,8, TCDD TOXIC EQUIVALENCIES
FOR DIOXINS AND FURANS

Concentration in
Nanograms per Dry
Standard
Cubic Meter
Corrected to 12% CO_2

	U.S. EPA METHOD	EADON METHOD
PCDD	0.0273	0.0335
PCDF	0.0127	0.0377
TOTAL PCDD+PCDF	0.0400	0.0712

of Air Quality Planning and Standards at Research Triangle Park had expressed an interest in conducting emissions tests at MERC to gather data on modern RDF-fired facilities. So, in addition to the tests required by DEP, EPA concurrently conducted testing of the emissions from the Unit A boiler train during the dioxin stack testing. Parameters sampled and the sampling points of the EPA-sponsored tests are shown in the matrix in Table 6. Preliminary results confirm the low emission test results achieved at the stack (see Table 7.)

MERC ASH ANALYSIS

The Site Location Permit issued by the Maine DEP when the plant was first licensed required that the three streams of ash (bottom, fly, and spent scrubber reagent/reaction products) produced at the MERC plant be analyzed for toxicity. The license required that the following metals be tested in accordance with

TABLE 6. EPA/MERC TEST PROGRAM

SAMPLING LOCATION	PARAMETER
Spray Dryer Inlet	Particulate, O_2, CO_2, CO, SO_2, THC, HCl, Metals (Cd, Cr, As, Pb, Hg), PCDD/PCDF (a)
Spray Dryer Outlet	O_2, SO_2, CO_2, HCl
Fabric Filter Outlet	Particulate, O_2, CO_2, CO, SO_2, THC, HCl, Metals (Cd, Cr, As, Pb, Hg), PCDD/PCDF(b)
Ash Discharges (Bottom Ash & Fabric Filter)	Metals (Cd, Cr, As, Pb, Hg), % Combustibles, % Carbon
Spray Dryer Inlet Lime Slurry	Metals (Cd, Cr, As, Hg, Pb)

(a) Separate analysis of front and back half.
(b) Combined front- and back- half analysis.

TABLE 7. PRELIMINARY RESULTS OF EPA TESTS AT MERC (BOILER A)

HCl	15 ppm
SO_2	21 ppm
VOC	1.1 ppm
CO	74 ppm
NO_x	207 ppm
TSP	0.01 gr/dscf
PCDD	1.3 ng/dscm
PCDF	2.7 ng/dscm

the EP Leachate Toxicity test: Arsonic, Barium Cadmium, Chromium, Lead, Selenium, and Silver. Additionally, the following elements were to be tested as total metals in the ash stream: Aluminium, Arsenic, Barium, Boron, Cadmium, Calcium, Chromium, Hexavalent, Chromium (Hexavalent and Total), Copper, Iron, Lead, Nickel, Magnesium, Maganese, Mercury, Molybdenum, Potassium, Selenium, Silver, Sodium, Vanadium, and Zinc. Once the plant design was finalized, it was recognized that only two separate ash streams (fly and bottom) would actually be produced. The spent scrubber reagent produced in the dry scrubber was actually being carried by the flue gas into the baghouse where it was collected and conveyed with the fly ash to the conditioning screw, mixed with the bottom ash, and then conveyed into trucks for disposal. The permit was modified to reflect this development and the combined ash/ fly ash/ scrubber product stream was included in the sampling and analysis plan. The DEP agreed that only the results of the combined ash stream characterization would be used in determining the non-hazardous/hazardous nature of the ash.

MERC was required in the license to analyze the ash for three (3) consecutive months after startup. Utilizing only four grab samples per month as outlined in the permit revealed questionable and often unreliable data. Presented with questionable results, MERC consulted with several private consultants to assist in designing an ash sampling protocol that would reliably characterize the ash.

The protocol, which was submitted to the DEP and approved, required that the facility up-grade its sampling requirements substantially. Under the new protocol, every four (4) hours, a representative one-pint sample was taken of each ash stream: fly ash, bottom ash, and combined ash. At the end of each 24-hour period, the six (6) one-pint samples of each ash stream were thoroughly mixed and two (2) one-pint samples representing each stream were then taken. At the end of either a three (3) or four (4) day cycle, one of each daily sample of each stream were thoroughly mixed and two (2) one-pint samples representing each stream were taken. One semi-weekly sample was then sent out for analysis. The unused daily and semi-weekly samples were archived in the event they were needed for further analysis. The results of the analyses of the combined ash stream during the next seven months of operations revealed consistent results well below the regulatory limits (see Table 8).

Initially required by the permit to provide ash analysis data for three consecutive months, the DEP insisted that sampling and ash characterization continue until at least 30 days after the initiation of commercial operation which began on December 29, 1987. The MERC plant began to generate ash on a regular basis in July 1987 and for the next seven months ash samples were composited and anaylzed on a regular semi-weekly basis. After five months of sampling with this new protocol, MERC received approval from DEP to lessen the daily sampling frequency from one sample every four (4) hours to one sample every six (6) hours. No change in sampling accuracy was noticed based upon the analytical test results. DEP has recently requested that MERC continue its ash characterization program on a quarterly basis.

The EPA procedure (SW846) for determining whether a solid waste such as the MERC ash is non-hazardous requires that the 90% upper confidence level (one-sided) of the sampled population be below the EP Leachate regulatory threshold limit for each of the nine heavy metals. Throughout the entire sampling period, utilizing the revised sampling protocol, the MERC facility consistently produced a final ash product that was non-hazardous and suitable for landfill disposal.

CONCLUSION

The results of the emission tests and ash sampling and analysis program at the MERC facility demonstrate that a modern RDF-fired facility equipped with an acid gas removal system and a high efficiency particulate collection device can effectively minimize the emissions of acid gases, heavy metals, trace organic compounds, and particulates, while producing a residue which is non-hazardous.

TABLE 8. MAINE ENERGY RECOVERY COMPANY
EP LEACHATE ASH ANALYSES
(ppm)

	Average	Standard Deviation	Upper Confidence Level	Regulatory Threshold Limit
Arsenic	0.04	0.0	0.04	5.0
Barium	1.81	0.91	2.00	100.0
Cadmium	0.16	0.22	0.21	1.0
Chromium	0.10	0.01	0.10	5.0
Lead	0.93	1.06	1.16	5.0
Mercury	0.001	0.002	0.001	0.2
Selenium	0.04	0.0	0.04	1.0
Silver	0.10	0.01	0.10	5.0
Vanadium	1.03	0.16	1.06	200.0

n=38
pH range 10-12

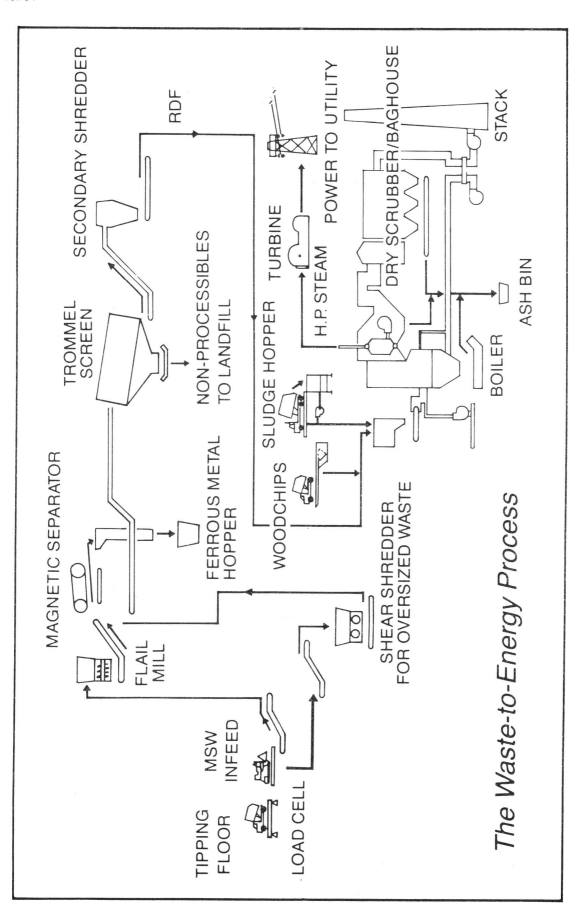

Figure 1. Solid waste.

COMBUSTIBLE MATERIALS RECOVERY AND MSW FLUIDIZED BED INCINERATION IN JAPAN

Takuya Kotani and Katsuhiko Mikawa ■ Mitsui Engineering & Shipbuilding Co., Ltd., Tokyo, Japan

This paper introduces the current trend of MSW processing in Japan. A recently constructed waste shredding and separating plant (1,250 metric tons/day) which recovers combustible materials, and fluidized bed MSW incinerators.

INTRODUCTION

In Japan, industrial waste discharged from factories and other industrial establishments must be properly treated by the private enterprise concerned. General waste resulting from everyday life must be treated by local government in accordance with the Governmental Law.

For waste discarded by households, which makes up the greater part of general waste, local governments are supposed to set up long-term processing plans, collect the waste before problems occur in the neighborhood environments, then transport and process it. Due to the rise in the standard of living and increased social and economic activities, the types of waste have become diversified and its quantity has increased to 120,000 tons per day, 1.4 times that of 10 years ago. This has created problems securing landfill sites necessary for processing this waste.

Population density in Japan is 320 people per km^2, or 12.5 times that in the USA. The world's greatest industrial production facilities are concentrated in this restricted area, and some 120 million people live there. Waste production is also so great that it is almost unrivaled anywhere in the world. Japan must dispose of this great amount of waste. One of the common methods is incineration, and 70% of the general waste is disposed of by burning.

For the past twenty years, stoker-type incinerators have mostly been used. Recently, however, more and more fluidized bed incinerators are coming into use, and their use is expected to increase in the future.

The increased amount of waste makes it difficult to obtain sufficient land to be landfilled, and increases waste treatment costs. Above all, the rate of waste collection and transportation has become very great (about 60 percent of disposal cost).

For these reasons, local government must process waste, taking care not jeopardize safety in neighborhood environments, and they must reduce the amount of waste by collecting classified waste, recycling waste, and making effective use of waste while maintaining the required service level.

Classifying waste and selecting what can be recycled before waste is collected by local government makes a great contribution, not only for effectively using resources but also for reducing waste.

WASTE PROCESSING IN JAPAN

Municipal Solid Waste Quantity

The quantity of waste produced is proportional to the degree of economic activity. This seems to be true in that up to the beginning of the 1970s the quantity of waste increased rapidly, but with the energy crisis of 1973 the increase in the quantity of waste fell off and has been level in recent years. As of 1985, the total population in Japan was 121,267,000, with 120,774,000 people living within areas where waste is collected. The quantity of waste collected was 96,940 tons per day, and if waste which was directly brought in is included, the quantity of waste processed by local governments was 113,782 tons per day. When the amount of waste processed privately (5,259 tons/day) is added, the total amount of waste produced per day was 119,041 tons. One person produced about 1 kg of waste per day.

General household waste which is collected by local governments makes up about 80 percent of the total. Directly delivered waste, which is mainly from businesses, accounts for 15 percent of the total. Bulky waste accounts for approximately five percent.

Waste Processing Methods

The amount of waste is reduced and made safe through incineration. The ash which is produced from incineration is then landfilled. In 1985, 70.6% percent of the waste which was collected or delivered was incinerated. Through the construction of incineration facilities, the percentage of waste which is incinerated is gradually being increased.

Waste Processing Facilities

There are 1,936 waste processing facilities in Japan. Of these, the majority (1,888 facilities) are incinerating facilities. The total capacity of these facilities is 161,000 tons per day.

Processing Costs

On a national scale, processing costs total 1.01 trillion yen. When the amount of waste processed per day is calculated as 114,000 tons, this brings the cost of processing per ton to approximately 24,000 yen.

When costs are calculated on the basis of the city size, calculations show that larger cities have higher costs. Of the different steps in processing (collecting, intermediate processing, and final processing), collecting is the most expensive, averaging 65.9 percent of costs or far more than half of all waste processing costs.

Classified Collection

The various types of wastes produced by households are classified by the discharger before the collection stage in Japan. The usual types of classification are: combustible waste, so that the efficiency of incineration is increased; resource waste, so that resources can be effectively recycled (resource recovery classification); hazardous waste, so that environmental pollution can be prevented; etc. Depending on the conditions in a given region, waste is classified into 2 to 5 types.

Resource recovery classification not only leads to effective use of wastes and recovery of resources, but also contributes significantly to reducing waste which must be incinerated and used as landfill. There are two types of resource recovery. One is resource recovery by residents, and the other is resource recovery by local government. In the former, groups of residents collect old newspapers, bottles and other reusable waste, then a resource recovery dealer with which these groups have a tie-up gathers the waste for recycling. In the latter, the local government gathers waste classified by the resident, then recycles any resources included. Resource recovery by residents is especially effective in reducing the amount of waste.

In a certain city, resources were recovered from waste by the residents. The amount of waste collected by the city was reduced from 20,400 tons to 18,000 tons, showing a huge reduction in the amount of waste. The table 2 compares the figures before and after resource recovery.

UTILIZATION OF WASTE AS A RESOURCE IN JAPAN

Conventionally, municipal waste in Japan has not contained as much paper as in the west, especially in the United States, and the water content of waste in Japan is twice that of the west. This makes it difficult to utilize waste in the form of RDF, and even if RDF is produced, there are few facilities which can use RDF. This is the main reason why many RDF devices have not been installed.

The first oil crisis awakened us to the

fact that there is a limit to petroleum resources, and consequently, recovery of resources and energy sources from municipal waste is being called for. The Ministry of International Trade and Industry started a ten year project in 1972 called STARDUST 80, in which research and development in processing technologies to aggressively utilize municipal waste as a resource were promoted.

Currently, there are three commercial plants in operation which convert waste to RDF in Chiba, Aichi and Mie Prefectures. All three plants utilize resource recovery system technology from PLM Sellbergs AB of Sweden. Their capacities are 80 tons/day, 47 tons/day, and 27 tons/day.

In the past, all incombustible waste, including waste unfit for incineration, which is collected separately from combustible waste was used as landfill. In 1987, the Tokyo Incombustible Waste Processing Center was built to process such waste and use it as a resource.

INCOMBUSTIBLE WASTE SHREDDING AND SEPARATING PLANT

In Tokyo's 23 wards, municipal solid waste is classified into three types of waste by each discharger, combustible waste, incombustible waste including waste unfit for incineration (hereinafter called incombustible waste), and bulky waste. Combustible waste includes kitchen garbage, paper, clothing, etc. and is collected three times a week. Incombustible waste includes metals, glass, ceramics, plastics, leather, etc. and is collected once a week. Bulky waste includes electric appliances, furniture, bicycles, etc. and is collected twice a month upon request. The population of the 23 wards is approximately 8.34 million and in 1987 the amount of combustible waste per day was expected to be 10,553 tons, incombustible waste 3517 tons, and bulky waste 410 tons. Fifty tons of combustible waste is sent to the compost processing facility, where three tons of compost is produced. The remaining 10,503 tons is sent to the 13 processing plants where it is incinerated. The 410 tons of bulky waste is sent to the shredding processing plant where it is processed.

Conventionally, all incombustible waste was used as landfill. The Incombustible Waste Processing Center was constructed to help effectively utilize landfill sites for long periods and utilize waste as a resource. It is designed to shred 1250 tons of incombustible waste per day. At the processing center, there is a plant to shred, separate and recover waste; a plant to melt and granulate plastics such as film, styrofoam, etc., into pellets; and a plant to collect and store landfill gas for power generation on site. The capacity for processing plastics is 100 tons/day (solids produced). Plastics are pelletized with a granulation agent in a rotary kiln heated by hot gas.

The fuel for the three power generator engines is landfill gas produced from nearby landfill sites, and the generating capacity is 320 kW each. Exhaust gas from the power generation units is used as a heat source for plastic processing facilities.

The incombustible waste shredding and separating plant which was constructed by Mitsui Engineering & Shipbuilding Co., Ltd. and has started operation in 1987, is described below.

Plant Flow Diagram

Figure 2 shows the process flow diagram. This plant shreds incombustible waste, then separates it into ferrous material, aggregate, plastic film, and other material containing much combustible material (hereinafter called combustible material), so that waste can be utilized as a resource and life of the landfill area will be extended. The total processing capacity of the shredding and separating plant is 1,250 tons/day (19 hours operation/day). The plant consists of two trains.

The plant is designed to recover about 200 tons of ferrous material, about 80 tons of plastic films, about 450 tons of aggregates, and about 520 tons of combustible material from 1250 tons of incombustible waste.

Receiving and Supplying Equipment

Incombustible waste is collected from each house by collection vehicles, then loaded aboard barges from the collection vehicles at the depot, and delivered to the dock near the center. There it is transferred from the barges to dump trucks (capacity 18 m^3) by cranes. After the waste in the trucks is measured on the scales, it is taken to the storage yard and unloaded. Waste which is stored is placed in the receiving chute with the shovel loader (8 m³ bucket). It travels on the receiving conveyor and feeding conveyor, and is fed to the shredder. An operator monitors the waste remotely and directly for dangerous waste or waste not suitable for the shredding process and removes any such waste.

Waste scales (1)
    Type   : Load cell
    Capacity   : 20 tons
Receiving chutes (2)
    Width   : 4 m
    Length   : 20 m
Receiving conveyors (2)
    Type   : Apron
    Width   : 2.2 m
Feeding conveyors (2)
    Type   : Apron
    Width   : 2.2 m
Unsuitable material removal devices (2)
    Type   : Hydraulic boom crane
    Capacity   : 300 kg

Shredding Equipment

Waste is continuously loaded into the shredder from the feeding conveyor. The shredder is of Mitsui-Williams horizontal swing hammer type, and shreds waste through impact force. Waste is dropped from a high inlet directly onto the hammer unit which rotates at 900 rpm. This is the initial shredding stage through impact. Roughly shredded waste is fed downwards between the impact blocks by the hammer. There is only a small gap between the impact blocks and hammer, so fine shredding occurs here. The finely shredded waste is sent further down to the cage bar for further shredding. Compression is repeated making the waste finer. The fine shredded waste is then discharged through the holes in the cage bar. Waste which has been thoroughly shredded falls down to the discharge conveyor. Waste which did not pass through the cage bar is lifted towards the casing top at a high speed by the hammer. This waste makes impact with the waste which is being fed from the feeding conveyor, causing further shredding, then all the waste falls down to the hammer unit. Because this system shreds waste through impact, shearing and compression, it is suited to shred incombustible waste containing various kinds of waste. The features of this shredder unit are as follows:

- Because waste is loaded directly from the conveyor without a compression feeder, various types of irregularly shaped waste can be loaded without limit.

- A 25 mm steel plate is used in the casing, creating a structure which can withstand the impact of the shredder.

- A hydraulic unit allows automatic centralized lubrication of the main shaft bearings, and the inspection cover can be opened and closed through a single action by a hydraulic cylinder.

- A special steel alloy is used in the hammer, and because the hammer is a hinged swing type, it can be attached in any angle.

- Because the unit has a rotor which can be rotated in both directions, both sides can be used uniformly and for long periods without hardfacing of the hammer.

- Because the pair of impact blocks can be adjusted with the hydraulic unit, the shredding size can be adjusted. This allows the gap between the hammer and impact blocks to be maintained at a constant width, even when the hammer is worn.

Shredders (2)
    Type   : Mitsui-Williams R-6200, 900 kW
    Capacity : 625 tons/19 hours
No. 1 shredded waste discharge conveyors (2)
    Type   : Vibration
No. 2 shredded waste discharge conveyors (2)
    Type   : Belt

Ferrous Material Separating Unit

The No. 1 magnetic separator at the end of the shredded waste discharge conveyor separates ferrous material from the shredded waste. At the outlet of the No. 1 magnetic separator is a ferrous material air classifier which separates light waste, and any metal which is discharged from the air classifier bottom is fed to the ferrous material sieve. This leads to recovery of high purity ferrous material. Ferrous material included in the unsifted waste through the sieve is recovered by the No. 2 magnetic separator.

No. 1 Magnetic separators (2)
    Type   : Electro and permanent magnetic
Ferrous material air classifiers (2)
Ferrous material sieves (2)
    Type   : Vibration

Area : 7.4 m^2
No. 2 magnetic separators (2)
   Type : Electro magnetic

Aggregate Separating Unit

After removal of ferrous materials shredded waste is sent to the medium mesh trommel, then sifted waste is sent to the fine mesh sieve. At the fine mesh sieve, fine aggregates are sifted out, and any waste which is not sifted is separated into coarse aggregate and light waste by the aggregate air classifier. Fine aggregate which is sifted through the fine mesh sieve is fed to the granulation agent sieve, and the fine, sand-like granulation agent is sent to the plastic processing device.

Medium mesh trommels (2)
   Diameter : 2.8 m
   Length : 6 m
Fine mesh sieves (2)
   Type : Vibration
   Area : 8 m^2
Aggregate air classifiers (2)
   Air velocity : 20 m/s
Granulation agent sieves (2)
   Type : Vibration
   Area : 3 m^2

Plastic-rich Waste Separating Unit

Waste which has not sifted out over the medium mesh trommel is fed to the large mesh sieve, and waste which also passes over the large mesh sieve is fed to the air classifier. Waste recovered by the air classifier is plastic-rich waste (mainly film) and is sent to the plastic processing device.

Large mesh sieves (2)
   Type : Vibration
   Area : 8 m^2
Plastic-rich waste air classifiers (2)
   Air velocity : 12 m/s

Combustible Waste

Waste left after ferrous material, aggregates, and plastics have been recovered contains much combustible material. Combustible material is mainly made up of paper, raw garbage, textiles, etc.

Transfer Units

Ferrous material which is recovered by the ferrous material sieve and No. 2 magnetic separator is fed to the ferrous material compacting unit by the ferrous material transfer conveyor. Aggregate which passes over the granulation agent sieve and discharges from the bottom of the aggregate air classifier is sent to the aggregate storage silo on the aggregate transfer conveyor. The sand-like granulation agent which was sifted by the granulation agent sieve is sent to the plastic processing device on the granulation agent transfer conveyor. Plastics are sent to the plastic processing device through the plastic air transfer unit. Combustible material is sent to the compactor through the combustible material transfer conveyor.

Ferrous material transfer conveyors (3)
   Type : Belt
   Width : 750 mm
Aggregate transfer conveyor (1)
   Type : Belt
   Width : 750 mm
Granulation agent transfer conveyor (1)
   Type : Belt
   Width : 450 mm
Plastic air transfer units (2)
   Duct diameter : 350 mm
   Air velocity : 150 m^3/min
Combustible material transfer conveyors (2)

Storage and Shipping Equipment

o Ferrous material
After recovered ferrous material is pressed into bales by the ferrous material compactor, it is stored in the bale storage yard. It is then loaded onto trucks with the bale storage yard crane and sold to scrap steel dealers.

o Aggregates
Aggregate and solidified plastics are stored in the aggregate storage silo and later shipped. Shipping is either from the shipping facilities or by the truck through the aggregate bunker.

o Combustible material
Combustible material is sent to the compactor and compressed into containers. The containers are stocked in the container yard. The containers are shipped out in container carrying trucks.

Ferrous material compactor (1)
   Type : Two-direction compression
   Main pressure : 400 tons
   Bale size (mm) : 600 (W) x 600 (H) x 700 (L)
Bale storage yard cranes (2)
   Type : Gantry crane with magnet

Rated load : 2 tons
Aggregate storage silos (2)
  Capacity : 350 m^3
Aggregate bunker (1)
  Capacity : 8 m^3
Combustibles material compactors (2)
  Thrust : 45 tons
Combustible material containers (85)
  Capacity : 18 m^3

MSW FLUIDIZED BED INCINERATION PLANT

In the history of MSW incinerators in Japan an 11.5 t/d batch incinerator was built in 1900 and the first continuous stoker type incinerator with a capacity of 18 t/d in 1960.

Today, 1,888 MSW incineration plants are being operated, and total incineration capacity is 147,473 tons/day.

An MSW incinerator with a power generator appeared in 1964, and about one hundred (100) MSW incineration plants are now generating electric power. Though stoker type incinerators have been mainly used for the past 20 years to burn MSW, the fluidized bed incinerators have advantages over the stoker type in treating hazardous-material- containing waste and in meeting stringent environmental requirements.

The first prototype of a fluidized bed incinerator was constructed in 1966. A fluidized bed has long been in use in process plants for gasifying coal, refining petroleum, roasting pyrites, baking cement, etc. It is now also being used in incinerating sewage sludge, plastics, waste tires, etc.

As customers realize the excellent performance, about 90 incineration plants equipped with a fluidized bed incinerator have been or are being constructed. The waste treatment capabilities have reached about 10,000 tons/day.

The number of fluidized bed type incinerators to be constructed has been gradually increasing.

Features of Fluidized Bed Incineration Plant

Features of the fluidized bed incinerator can be compared with those of stoker type incinerator as follows:

o The fluidized bed incinerator can effectively burn all types of waste under stable conditions, from moist waste to high calorie waste containing plastic and rubber. It even accepts slurry sewage sludge and normal MSW simultaneously.

o Because of fast combustion, the response for waste quality and quantity is very quick, and stable operation under optimum conditions can easily be obtained.

o Combustion is carried out under stirring fluidized condition so that local high heat combustion, refractory degradation by heat, and clinker generation are avoided.

o The fluidized bed incinerator has no mechanical moving parts. The simple and sturdy design ensures extended maintenance-free operation.

o Combustibles are burned completely and the discharged dry residue contains only combustibles of less than 1%, so that the residue treatment cost is very low and the service life of the landfill site is extended.

o Because of the compact furnace and large heat content of the fluidizing material, starting and stopping become very easy.

The fluidized bed incinerator is suitable for continuous operation as well as semi-continuous operation (for example, 8 or 16 hour operation a day) which requires start and stop operations everyday.

MSW Fluidized Bed Incinerator

Figure 3 shows a typical MSW fluidized bed incinerator. Incineration is carried out as follows:

o Fluidizing air is fed into an incinerator to form the fluidized bed by a primary air blower through air nozzles.

o The fluidized bed is heated and maintained at high temperature by a heat-up burner.

o The heat-up burner is turned off when the fluidized bed is heated to the temperature at which the waste can be burned.

o As soon as waste is spread by the spreader in the high temperature fluidized bed incinerator, the waste is mixed and abraded with high temperature sand, and then dried, and swiftly burned. The waste is completely burned in the high temperature freeboard zone.

o Incombustibles in the waste are discharged with fluidized bed materials through a bed material discharger installed at the bottom of the incinerator. They are then separated from the fluidized bed materials by a separating screen. Bed materials are returned to the incinerator by the bed material conveyor.

o Flue gas from the freeboard of the incinerator is discharged into the atmosphere through a waste heat boiler (or gas cooling tower), an electrostatic precipitator, etc., by a forced draft fan.

The fluidized bed temperature is controlled over 650°C. If moist waste of low calorific value is fed into the incinerator, auxiliary fuel is necessary to assist incineration.

o The fluidized bed incinerator is normally capable of burning 3,100 kJ/kg waste without any auxiliary fuel. If fluidizing air is heated to high temperature by hot flue gas, 2,500 kJ/kg of waste can be burned without auxiliary fuel.

The incinerator is designed to be very simple, vertically cylindrical or rectangular, and is made up of refractory materials and steel wall.
The combustion chamber is divided into four parts: freeboard, fluidizing bed, air injection, and incombustibles outlet, from top to bottom.

The waste spreader, the heat-up burner and the secondary air nozzles are provided about half way of the incinerator.

The fluidizing part is filled with sand for fluidization to a depth of 500 to 1000 mm in the stop condition. In normal firing condition, the fluidized bed is formed by injecting air from the air nozzles.

Waste Quality

The waste quality differs, depending on the season, climate, region, economic and social conditions, etc. Table 4 shows an example of waste quality in Japan (an urban area) and USA (a county in a southern state).

Outline of MSW Incineration System

Figure 4 shows the system flow of an MSW fluidized bed incineration plant.

Waste is collected and brought to the incineration plant by the collection vehicle, and weighed on the weigh-in scale, which automatically measures the weight of waste. The layout of the truck road is designed so that the feed trucks are not disturbed by outgoing trucks.

Hatches for the waste pit are kept shut except when the waste in the collection vehicles is dumped into the waste pit by gravity.

The primary and secondary air blowers induce air into the waste pit to prevent odors from being dispersed into the surrounding area, because waste in the waste pit emits offensive odors. The primary air is discharged through the air preheater to the incinerator to burn the waste. Air is preheated to 100 to 250°C depending upon calorific value of the waste and injected through the air nozzle in the incinerator. Secondary air is discharged to the freeboard part in the incinerator as excess air.

Bulky waste, such as timber, mattresses, wooden furniture, etc., is first cut into pieces by a guillotine cutter, and then discharged to the waste pit.

Waste in a waste pit is fed to a waste feeder by a waste crane used for feeding and mixing waste in the waste pit, and is then spread into the incinerator.

The fluidized bed temperature is kept over 650°C, so waste begins to burn instantly and burning is completed in the freeboard section.

Flue gas from the incinerator is cooled to about 300°C by the gas cooling tower or the wasteheat boiler and fed to the electrostatic precipitator or bag filter to remove ash from the flue gas. The gas is discharged into the atmosphere through the stack by a forced draft fan, the speed of which is controlled according to the load, to reduce electric power.

Steam generated by the wasteheat boiler is utilized for the steam turbine generator, which supplies electric power to the various machines in the plant. Surplus electric power is distributed to the electricity network. Steam is also used for cogeneration plants where high-pressure steam generates electric power, and low-pressure steam from the generator turbine is supplied for district heating.

Incombustibles in waste are collected from the bottom of the incinerator as combustion residue and from the gas cooling tower or the wasteheat boilers and the electrostatic precipitator as ash. Ferrous material in the combustion residue is recycled by a magnetic separator, and other materials are used for landfill with ash, which is solidified.

Pollution Control

Dust. The dust contained in the exhaust gas is removed in two ways: large particles of dust are collected by gravity in the cooling tower or the wasteheat boiler, and fine particles by the electrostatic precipitator. Purified flue gas containing dust of less than 50 mg/Nm^3 is discharged into the atmosphere.

Nitrogen oxides (NOx). Waste combustion produces nitrogen oxides in two forms; fuel NOx produced by the nitrogens in the waste, and thermal NOx produced by those in the combustion air. Since a low excess air ratio is applied for the fluidized bed incinerator, the overall NOx generation is low 150 ppm. A two-stage combustion system can reduce the NOx to less than 100 ppm.

Hydrogen chlorides (HCl) and sulfur oxides (SOx). Hydrogen chlorides generated by the combustion of petrochemical products in the waste are corrosive and harmful so that HCl removal becomes very important.

A dry type HCl removal process (below 250 ppm), semi-wet type (below 100 ppm), or wet type (below 20 ppm) is used to meet stricter emission standards.

Incombustible residue. The fluidized bed incinerator can completely burn all types of waste, from high calorie waste containing plastic to sewage sludge, so that there are only incombustibles such as dirt, glass, shards, and metal in the residue discharged from the bottom of the incinerator. Residue free of combustibles is used for landfill without any treatment after metal recovery.

Ash with ignition loss of less than 1% is collected in the cooling tower or the wasteheat boiler and the electrostatic precipitator in a dry condition. It is used for landfill after wet conditioning, or after cementing into blocks to prevent heavy metal from leaching out and to ensure bearing pressure in the landfill area.

Wastewater. Pit slops are heat-treated in the incinerator for oxidization. The wastewater from the plant is treated and re-used within the plant.

Foul odors. The foul odors in the waste pit are sucked into the incinerator and decomposed at high temperature. The entire system is designed to prevent foul odors from being emitted into the surrounding atmosphere.

Noise and vibration. Machines that generate noise and vibration such as the air blower, the air compressor, and the shredder are installed in an independent space.

Operating Results

Operating results of a 40 ton/16 hour municipal waste incineration plant are presented. The plant was tested under the conditions shown in Table 5, and its running temperatures are shown in Fig. 5.

Plant startup and shutdown. The plant is operated semi-continuously, working for 16 hours a day, and an operator must start it up and shut it down every day. When the fluidized bed temperature is 560 to 600°C, no auxiliary fuel is required for startup heating; the plant can be restarted if waste is charged directly. When the bed temperature is below 500°C, a bed burner must be used for the startup. When the bed temperature is between 500 and 560°C however, the incinerator can be heated to operating temperature by starting the bed burner and by charging waste at the same time, thus reducing heavy oil consumption and startup time.

Normally, startup takes about 15 minutes, and the time required to attain normal operating temperature of 700°C or higher is not more than 30 minutes.

To shut down the plant, the waste feed is cut off, and the combustion air blower and induced draft fan are turned off as soon as the incinerator monitor CRT and flue gas O_2 meter show that combustion has been completed. About 5 minutes is required for this shutdown operation. Other ancillary machines are then brought to a stop sequentially. The total time required for plant shutdown is about 30 minutes.

Temperature control. The temperature of the fluidized bed is held at 650°C and over, chiefly by regulating the waste feed rate. If the fluidized bed temperature falls below 600°C under a rated load, the bed burner is fired. In addition, the primary air temperature is regulated within 100 to 250°C to keep the optimum combustion condition for any quality of waste.

Ignition loss. The ignition loss of ash sampled from the gas cooling tower and the electrostatic precipitator during plant operation was less than 1%. Ignition loss was measured by roasting the ashes at 600° + 25°C for 3 hours.

Emission analysis. Under a rated load, the emission characteristics met the emission standards satisfactorily and were as follows:

Dust concentration	0.032 g/Nm^3 (DG)
Sulfur oxides (SOx)	14.9 ppm
Nitrogen oxides (NOx)	106 ppm
Hydrogen chloride (HCl)	178 ppm

Table 1. Municipal waste processing costs.

City population	Processing cost (yen/ton)	Processing cost breakdown (%)		
		Collecting	Intermediate processing	Final processing
1,000,000 or more	23,200	67	25	8
More than 300,000, less than 1,000,000	20,000	54	35	11
More than 100,000, less than 300,000	14,000	57	37	6
Less than 100,000	13,500	57	36	7

Table 2. Example of resource recovery.

	Before resource recovery	After resource recovery
Amount of waste (ton/year)		
Combustibles	16,600	15,800
Incombustibles	3,800	1,000
Total	20,400	16,800
Per head (kg/year)	241	211
Recovered Resource (t/year)	-	3,200
Final disposal quantity (t/year)	5,400	2,981
Final disposal cost (Mil.yen/year)	108.1	29.5
Subsidy (Mil.yen/year)	1.7	17.4

Table 3. MSW incineration plants in Japan.

Type	No. of plants	Total capacity (t/d)	Average capacity (t/d)
Fixed bed (batch)	288	3,236	11
Stoker (batch)	1,014	28,716	28
Stoker or fluidized bed (8 or 16 hr operation/day)	198	15,448	78
Stoker or fluidized bed (24 hr operation/day)	388	110,073	284
Total	1,888	147,473	78

Table 4. Waste quality.

		Japan	USA
Combustibles	wt %	43.5	52.98
Carbon		(21.46)	(25.6)
Hydrogen		(3.05)	(3.45)
Oxygen		(18.06)	(23.0)
Nitrogen		(0.66)	(0.60)
Sulfur		(0.06)	(0.17)
Chlorine		(0.21)	(0.16)
Ash	wt %	10.4	21.52
Water	wt %	46.1	25.50
Lower calorific value	kJ/kg	7,100	9,800

Table 5. Operating data.

		Test 1	Test 2	Test 3
Feed rate	kg/h	2,510	2,730	2,550
Calorific value	kJ/kg	7,200	5,020	7,370
Bed load	kg/m^2·h	423	460	430
Bed temp.	°C	700	680	730
Incinerator outlet temp.	°C	850	760	880
Primary air temp.	°C	200	250	180
Auxiliary fuel	/h	0	0	0
Fluidizing air	Nm^3/h	4,700	4,700	4,700
Secondary air	Nm^3/h	3,100	2,300	3,200

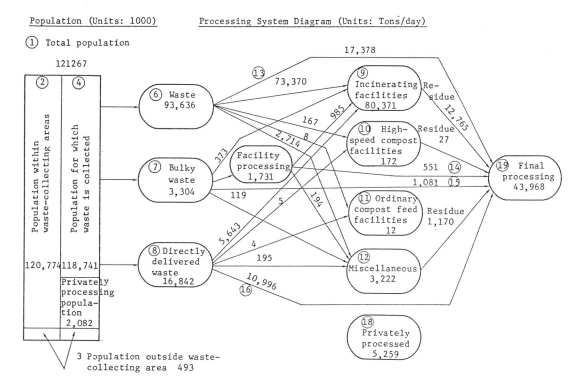

Figure 1. Municipal solid waste processing in Japan (1985).

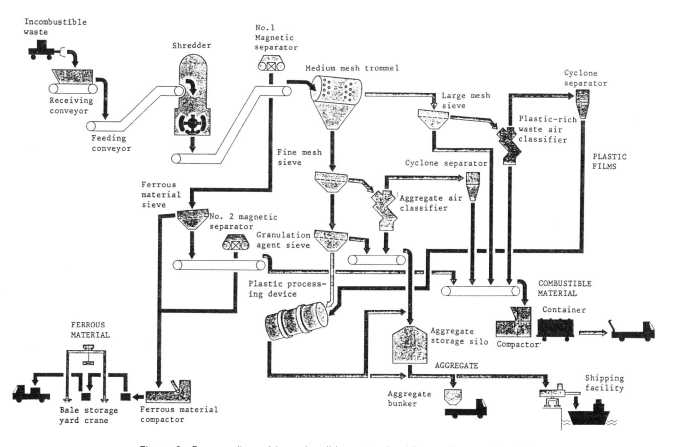

Figure 2. Process flow of incombustible waste shredding and separating plant.

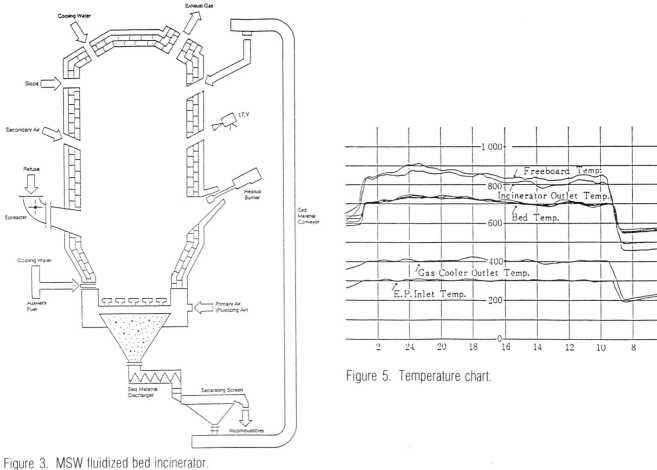

Figure 3. MSW fluidized bed incinerator.

Figure 5. Temperature chart.

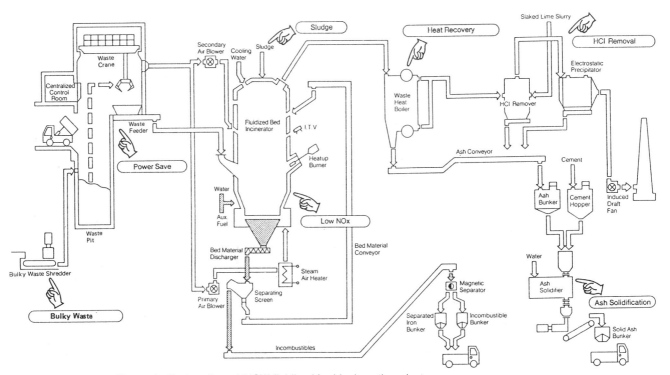

Figure 4. System flow of MSW fluidized bed incineration plant.

AN ITALIAN PROCESS FOR HANDLING MUNICIPAL SOLID WASTES

William J. Sim ■ American Recovery Corporation, Suite 8350, 2000 Pennsylvania Avenue, N.W., Washington, DC 20006

Sorain-Cecchini Sp.A., an Italian company based in Rome has been dedicated to the design, fabrication, and operation of Municipal Solid Waste (MSW) processing systems for over 25 years. Based on a continuous research and development program, Sorain provides the most state-of-the-art technology for processing MSW to recover recyclable materials from MSW. During the last 25 years, the Sorain process has handled over 15 million tons of Municipal Solid Waste. All Sorain facilities optimize the recovery of materials that have a reliable commercial market within the local community including paper, corrugated cardboard, aluminum, ferrous materials, and organic material for compost feedstock. One of Sorain's most unique capabilities is the fully automated film plastics recovery process which produces a high grade film plastic product that is used in the manufacture of plastic bags, pipe, and conduit.

In January 1988, American Recovery Corporation (ARC) was formed as a long-term Joint Venture between Sorain-Cecchini Sp.A. of Rome, Italy and Potomac Capital Investment (PCI) of Washington, D.C. U.S.A. This Joint Venture now represents the Sorain Municipal Waste Processing technology in the United States and Canada.

Potomac Capital Investment Corporation (PCI) is a wholly owned subsidiary of Potomac Electric Power Company (PEPCO) and conducts PEPCO's non-utility investment program including interests in small power production which PCI carries out through its subsidiary known as American Energy Corporation (AEC). Currently this small power production program includes ownership in three 30 MW solar electric generating stations located in California's Mojave Desert producing power for sale to Southern California Edison. AEC is also actively pursuing hydro, cogeneration, and other small power production projects in the United States.

The purpose of this paper is to present the Sorain technology, its experience in Europe, and how it uniquely suited to applications in the U.S. resource recovery marketplace. In fulfilling this objective, the paper presents:

(i) Background - The past accomplishments of Sorain-Cecchini;
(ii) Why materials recovery;
(iii) A description of the technology; and,
(iv) How the technology can be applied in the U.S.

BACKGROUND

The history of Sorain-Cecchini dates to 1942 when the Cecchini Company was founded as a waste collection company in Rome. In 1976, Cecchini merged with two other companies, Sarr (founded in 1947) and S.O.R.A.I.N. (founded in 1956) to form Sorain-Cecchini, Sp.A. as it is known today. Each of the merging corporations collected and recycled residential refuse in the City of Rome and each operated a recovery facility within the city limits. The plants were built in 1962, 1964, and 1965 and a total of 2,000 tons per day was processed on a regular basis. With the merger, the resource recovery experience, methodologies, and staff of the three companies were combined.

In the twenty-five years since the first Rome plant went on line, Sorain systems have processed over 15,000,000 tons of solid waste. Since 1976, Sorain has standardized their processing methods, developed product enhancement technologies, devoted a significant effort to research and development activities, and expanded their technology application beyond Italy to other European countries, South America, and now to North America.

The Sorain technology is utilized in the seven operating plants listed in Table 1. The extent of Sorain's experience is best expressed through a technical description of these facilities and their involvement in each, which are presented in the Appendix of this paper.

Table 1. Sorain-Cecchini Operating Plant

Location	First Year of Operation
Rome West, Italy	1964
Rome East, Italy	1967
Pomezia I, Italy	1977
Pomezia II, Italy	1979
Perugia, Italy	1972
Rio de Janeiro, Brazil	1975
Oslo, Norway	1985

In addition, Sorain has three plants under various stages of design and construction (see Table II). These plants and Sorain's involvement in each are also described in the Appendix.

Table II, Sorain-Cecchini Plants Under Construction

Location	Scheduled Operating Date
Prague, Czechoslovakia	Early 1988
Cassino, Italy	Early 1988
Milan, Italy	Early 1989

Through this design and operating experience, Sorain has developed and refined the unique ability to mechanically separate mixed paper, ferrous and film plastic, and regenerate film plastic to produce a granulate suitable for use in product manufacturing. In addition, refinements to normal processing techniques have been made to further enhance the quality of materials and refuse-derived fuel recovered using Sorain systems.

WHY MATERIALS RECOVERY

The recycling of materials from the municipal solid waste stream has been practiced for decades. Recycling in the U.S.A. has had little direct impact on the amount of waste disposed in landfills or waste-to-energy facilities because economics have not driven the expansion of this effort beyond token, symbolic programs. These economics are driven by the availability of markets for the recovered (secondary) materials.

As a result of the global conditions of increasing population and decreasing reserves of natural resources available to produce virgin materials, the economics of secondary materials markets have demonstrated improvement over the past several years. For example, since 1979 the decline of worldwide timber reserves has created an accelerating long-term fiber shortage. A vivid indication of this trend is the 100 percent increase in the Pacific coast export of secondary fiber material (corrugated and newsprint) from 1977 to 1985. This quantity is expected to double again by 1989.

Such long-term trends indicate a general, continual strengthening of the economics of secondary materials recovery. In addition, an active and vocal sector of the populace advocates the maximum re-use of materials in the waste stream. The reasoning of this sector is that long-term environmental and natural resource trends logically indicate the desirability and necessity of establishing a policy to recover and re-use materials. This re-use encompasses recycling usable materials such as aluminum, ferrous, and paper as well as composting of the organic waste fraction.

The incorporation of material recovery into medium and large-scale resource recovery projects yields significant benefits to both the public and private sector, including:

- Flexibility. A materials recovery system has the flexibility to respond to changing materials and energy market conditions. For example, a recovery system can be modified to recover additional materials as markets develop, if appropriately designed. Also, if energy/fuel demand is cyclical, recovery rates of marketable combustible materials can be increased when energy demand is low and vice versa.

- Public Appeal. By presenting a total waste disposal solution to the public which (I) maximizes materials re-use (source separation augmented by mechanical materials recovery system), (II) incorporates energy generation, and (III) minimizes the quantity of waste landfilled, a broad coalition of public support can be assembled. This coalition may well include recyclers and environmental groups normally opposed to incineration or landfill projects.

- Capability with Combustion Technologies. Materials recovery systems can be compatible with mass burn, spreader stoker, or fluidized bed combustion technologies. With mass burn systems, materials can be recovered and the unshredded combustible fraction burned with a reliable and proven mass burn grate system. With a spreader stoker boiler, the combustible fraction can be sized using a high speed horizontal or vertical shaft shredder prior to combustion. Size reduction (and densification in some cases) will allow a prepared refuse fuel to be burned in a fluidized bed boiler.

- Air Emissions Enhancement. With processing of waste prior to burning, the volume of air emissions is decreasing in relation to the amount of waste accepted by the facility. This occurs for two reasons: First, depending on the waste composition and the types and qualities of materials recovered, only 50 to 70 percent of the incoming waste is burned. Therefore, a smaller boiler can be utilized with resulting lower flue-gas volumes than with a boiler burning 100% of the incoming waste. Secondly, through the removal of non-combustibles (i.e., ferrous metals, aluminum, glass and grit), wet organic waste, and other undesirable combustion materials, the resulting prepared fuel is a cleaner, more uniform, and higher BTU content fuel than unprocessed waste. Therefore, excess air requirements will be lower and boiler efficiency correspondingly higher. Also, emissions quality may be improved by the removal of certain problem constituents, such as plastics and metals in the processing system.

- Minimized Landfilling. With maximum recovery of materials, burning of the prepared fuel and composting of the process residue and yard waste, only ash, grit, and bulky non-processible waste must be landfilled. Therefore, the amount landfilled could be as low as 12 percent by weight (3 percent by volume) of the incoming waste.

- Ash Quantity and Quality Enhancement. With continuing

uncertainty regarding the non-hazardous designation of ash from refuse-to-energy facilities, it is important to note that smaller quantities of bottom and flyash will naturally result from smaller amounts of waste combusted in prepared burn systems. Also, ash quality can be enhanced through the removal or problem constituents in the processing system.

o <u>Project Economic Impact</u>. Materials recovery can enhance project economics. The positive economic impact of (I) lower boiler capital cost by virtue of smaller required size, (II) increased boiler efficiency, and (III) recovered materials revenues in many cases more than offsets any increased capital cost, operating cost, and in-house power consumption resulting from the incorporation of a processing system in a resource recovery facility.

GENERAL DESCRIPTION OF SORAIN-CECCHINI TECHNOLOGIES

The design of a recycle plant varies based on the following basic parameters:

(1) Waste Compostion;
(2) Materials to be recovered;
(3) Degree of recovery; and,
(4) Availability of a domestic or international market for the recovered materials.

All plants feature a Waste Receiving and Primary Sorting Section while the other auxiliary systems and satellite refining units are included case by case according to the function of the above listed parameters.

In a hypothetical maximum configuration, the recycling plant will consist of the following sections:

(1) Waste Receiving and Storage

(2) Primary Sorting

(3) Plastic Film (Ployethylene) Recovery

(4) Corrugated and Mixed Paper Recovery

(5) Aluminum Recovery

(6) Recovery of Mixed Organic Fraction

(7) Ferrous Recovery

(8) Batteries

(9) Recovery of Combustible Fraction (if required)

and the following further processing and refining units may be added:

(10) a. Composting of the Mixed Organic Fraction
b. Refining of the Compost

(11) Processing of the Mixed Paper Fraction into Paper Pulp

(12) Processing the Plastic Film Fraction into Plastic Granulate

(13) Refining and Densification of the Ferrous Fraction

(14) Densification of Aluminum Fraction

(15) Baling of the Corrugated Paper Fraction

(16) Processing of the Combustible Fraction into Refuse Derived Fuel (if required)

The end uses of the recovered materials listed above include the following:

o Reuse of the paper pulp and the corrugated in the paper industry

o Production of trash bags, pipe, and conduit from the plastic granulate

o Use of compost as soil conditioner in parks and gardens

o   Reuse of the ferrous in the steel industry

o   Reuse of the aluminum in the fabrication of beverage cans

o   Combustion of the refuse derived fuel and production of steam and electricity (if required)

Assuming the combustion of the refuse derived fuel, the residues of the process will be oversized and bulky (white goods, furniture, large tires, etc.) ash and rejects of the compost refining process (glass, ceramics, inerts, etc.) have undergone a sanitizing process as part of the composting process. These materials which represent on the average 20 percent by weight but only 10 percent by volume of the incoming waste, need to be landfilled.

GENERAL FLOW DESCRIPTION

All waste handling vehicles are weighed at the entrance to the plant and then directed to the unloading and storage area. Two configurations are possible:

(1) Closed pits serviced by a bridge crane and grapple system

(2) Closed floor level receiving area serviced by front loaders

The waste is loaded onto a steel plate apron conveyor which feeds the Primary Sorting Lines. The number of lines depends on the amount of waste processed daily and on the hours of operation of the facility. Processing lines are modular and different sized lines are available with rated hourly capacities ranging from 30 to 70 tons. Each infeed conveyor is serviced by an articulated crane, the function of which is to remove the waste stream oversized and undesirable items which are ultimately disposed of in the landfill. The latter, are discharged into an appropriate storage area or pit. The infeed conveyor is further equipped with an automatic Load Leveler which evens out the material flow into the processing line. The infeed conveyor carries the waste into the bag-breaker.

The bag-breaker is a low RPM machine thereby grossly reducing explosion hazards and has the function of opening up all waste containers and of sizing down large pieces of cardboard and other bulky items. This machine also displays the unique feature of automatically discharging into the reject line oversized and hard to break items. The combination of the pick-up crane and bag-breaker assures the absence of materials and items which may severally damage or clog the process machinery. The material leaving the bag-breaker continues on conveyors to a volumetric classification obtained by using rotating trommels equipped with specially designed self-cleaning grates which require little or no maintenance. In this stage, the mixed organic fraction is separated and conveyed to the composting process and the primary separation of ferrous takes place. The residual waste stream is then subjected to a stage of classification where we obtain, through the use of specially designed adjustable air classifiers, the separation of the list fraction (paper, plastic, film, textiles) from the rest of the material.

The light fraction is fed to a differential shredder, which distinzxguishes between the physical characteristics of paper and plastic film. This is obtained by exploiting the elasticity of plastic versus the fragility of paper. At the outlet of the differential shredder the size distributed of the plastic and paper fragments is such that it is possible to obtain a very high degree of separation through the use of a combination of rotating trommel and air classifier.

The heavy fraction of all the air classification stages together with the mixed fraction of the paper and plastic separation process can then be processed into refuse derived fuel, if required, by shredding material into a size range suitable for a high efficiency combustion.

The recovery of aluminum is achieved subsequent to the trommeling stage, while the corrugated and final ferrous recoveries are performed on the heavy fraction after the first stage of air classification.

The separated polyethylene film which still contains some contaminants in the form of other plastics, textiles, dirt, etc., can then be shredded and subsequently washed, dried, extruded and granulated. The final product can be blow extruded into film or used for the fabrication of plastic pipes.

The Sorain-Cecchini Group owns and operates, in the Rome area, facilities for the regranulation of the plastic film recovered from the MSW of Rome, as well as facilities for the filming of such recovered plastic resulting in the production of waste bags sold to Municipalities and consumers and the production of plastic pipes for irrigation and civil use.

The separated paper fraction is baled for intermediate storage and then can be fed into a paper pulping line.
The general characteristics of this line are the same as in a normal paper mill except for some added proprietary features necessary in order to process a paper fraction which, due to the fact that it is coming from waste, is accompanied by contaminates which are not normally present in normal scrap paper used in paper mills. The final product is a sanitized pump fibre ready for product refinement.

The ferrous fraction, consisting mainly of cans, is separated at different point along the process and then can be conveyed to a completely cold cleaning and densification process where the metal is stripped of labels, cleaned of contaminants, and densified into nuggets.

APPLICATION OF SORAIN-CECCHINI TECHNOLOGY IN THE UNITED STATES

The Sorain-Cecchini Technology is well suited to fit into total resource recovery system concept noted above. It's advantages include:

- o Long term experience in plant operation, design, and equipment fabrication;

- o The demonstrated ability to mechanically recover mixed paper and film plastic, ferrous, cardboard, and glass if required; and,

- o The demonstrated ability to enhance recovered materials, including the production of paper pulp, the regeneration of low density polyethylene to produce a granulate, and produce a high quality compost.

There are important practical issues which must be addressed by any foreign equipment fabricator or designer in adapting European technology to applications in the United States, as well as issues unique to the solid waste industry, that must also be addressed in such a technology transfer, including:

- o <u>Waste Composition</u>. The variation of waste composition in Sorain plant locations dictated that their design be flexible enough to easily accommodate those differences. In addition, we have adapted our design to achieve a variety of objectives in these different areas, such as recovering paper fiber, producing compost, fuel for energy recovery, paper pulp, and/or recovering film plastic.

Table III presents the composition of waste in selected Sorain plant locations and in Southern California.

Sorain-Cecchini adapted their Italian plant design, where waste is high in food waste and organics, to the Oslo installation the composition is more characteristic of U.S. solid waste. Therefore, design adaptation to the U.S. solid waste stream does not, in itself, present a new or significant technical challenge to Sorain.

- o <u>Adaption to U.S. Codes and Standards</u>. In it's European experience, Sorain designed and fabricated their equipment using Italian and German DIN standards. In general, the DIN standards for materials, fabrication, and electronic application are as stringent or more restricting than comparable U.S. standards (ASTM, NEMA, UL, etc.), thereby ensuring Sorain compliance with these standards.

Other codes and standards pertinent to system design are unique to the U.S., such as OSHA, EPA, Uniform Building Code, and the American Conference of Government Industrial Hygenists (re: dust collection and ventilation systems) requirements. These have been reviewed in detail by Sorain and ARC technical staff and are incorporated in our design.

o Spare Parts. Much of the equipment used in a Sorain processing system, such as belt conveyors, balers, horizontal shaft shredders, and dust collection equipment, are standard items procured from U.S. vendors to Sorain specifications. Therefore, spare parts for this equipment, which represents the majority of hardware in the system, are readily available. Any proprietary equipment fabricated by Sorain outside the U.S.A. are supplied with adequate spare parts. Motor mounts are provided with adaptor plates to ensure compatibility with U.S. motors or European motors carrying the UL label.

SUMMARY

There is a strengthening economic and political incentive in the U.S. to include the recovery and enhancement in resource recovery systems. American Recovery Corporation and Sorain-Cecchini have a technology and operating experience that offers a solution to this challenge. In particular, we have the ability to mechanically recover mixed paper and film plastic, and to enhance the quality of recovered materials, thereby increasing their marketability.

We have demonstrated an ability to transfer this technology to the United States by accommodating the U.S. waste composition, codes and standards, and capacity requirements. As a result of these efforts, large-scale materials recovery, with its inherent flexibility and associated advantages in conjunction with energy recovery, has been technically and economically proven.

APPENDIX
SORAIN-CECCHINI FACILITY DESCRIPTIONS

OPERATING FACILITIES

Rome West. Began operation in 1964; incorporates two 25 tons per hour (TPH) lines; recovers film plastic, ferrous metal, organics for composting, and mixed paper; Sorain responsible for system design, equipment fabrication, construction, and plant operations.

Rome East. Began operation in 1967; incorporates six 18 TPH lines, recovers film plastic, ferrous metal, organics for composting, and mixed paper; produces semi-densified RDF for scale to cement kiln; Sorain responsible for design, equipment fabrication, construction, and plant operations.

Pomezia, Italy. Began operations in 1977; regenerates 2500 tons per year (TPY) of film plastic recovered from Rome East and Rome West plants; granulate produced is used as a raw material in the manufacture of bags and pipe in an adjacent Sorain owned and operated factory; Sorain owns and was responsible for system design, equipment fabrication, construction, and plant operations.

Perugia, Italy. Began operations in 1972, expanded in 1980; incorporates one 25 TPH line and plastic regeneration system; recovers film plastic, organics for composting, mixed paper, and ferrous metal; produces densified RDF for sale to utility which co-fires RDF with coal; regenerates film plastic; Sorain responsible for system design, equipment fabrication, construction, and plant operations. Operations are conducted by Gesemu, a local Sorain subsidiary.

Rio de Janeiro, Brazil. Began operations in 1975; incorporates one 25 TPH line; recovers film plastic, cardboard, textiles, and organics for composting; Sorain responsible for system design and equipment fabrication.

Domtar Paper/Toronto, Canada. Began operations in 1983; recovers corrugated; Sorain responsible for system design and equipemnt fabrication.

Oslo, Norway. Began operations in 1985; incorporates two 28 TPH lines and plastic regeneration system; recovers mixed paper (for on-site pulping), film plastic, and ferrous metal; combustible fraction co-fired with unprocessed waste in Wider-Ernst mass burn furnaces; Sorain responsible for conceptual design for processing system and equipment fabrication/installation of the plastic regeneration system.

FACILITIES UNDER CONSTRUCTION

Prague, Czechoslovakia. Construction started in 1985, scheduled operations in early 1988; incorporates two 27.5 TPH lines, plastic regeneration system, and composting system; recovers film plastic, ferrous metal (cleaned and densified), organics for composting; produces semi-densified RDF for sale, regenerates film plastic, and composts organics; Sorain responsible for system design and equipment fabrication.

Cassino, Italy. Construction started in 1985, scheduled operations in early 1988; incorporates two 22 TPH lines, plastic regeneration system, and composting system; recovers film plastic, ferrous metal (cleaned and densified), organics for composting, mixed paper; will regenerate film plastic, produce paper pulp, and compost organics; Sorain responsible for design, equipment fabrication, construction, and operation.

Milan, Italy. Construction started in 1986, scheduled operations in early 1989; incorporates three 22 TPH lines, plastic regeneration system, and composting system; will recover film plastic, mixed paper, ferrous metal, organics for composting; will produce RDF for on-site boiler, regenerate film plastic, produce paper pulp, and compost organics; Sorain responsible for design, equipment fabrication/installation, and operations.

Table III

Waste Composition - Sorain-Ceechini Facility Locations

	Rome	Milan	San Paulo	Oslo	California
Paper/Cardboard	25.0	20.0	21.0	38.2	40.5
Film Plastic	3.5	5.0	2.6	4.7	2.0
Hard Plastic	3.0	5.0	1.7	1.8	5.4
Ferrous Metal	2.5	4.0	4.1	2.0	5.0
Textile/Leather/Wood	3.0	5.0	7.0	9.4	18.1
Organic Matter	53.0	41.0	57.0	30.4	19.6
Non-Combustibles (Glass, Non ferrous Metal, Sand, Grit, etc.)	10.0	10.0	6.6	13.5	9.4

PELLETIZING FOR WASTE UTILIZATION IN SWEDEN

Anders Larkert ■ PLM Sellbergs AB, Box 19605, S-10432 Stockholm, Sweden

PLM Sellbergs of Stockholm, Sweden, the main waste contractor in Scandinavia, has developed a process whereby household waste is converted into RDF, compost and scrap iron. The fuel component varies between 40 and 60% of incoming waste, depending on the composition of raw waste, desired quality of RDF etc. Typically 80% of the heat value of the waste is recovered by the process.

The technology being developed in the seventies, the first commercial plant was taken into operation in 1981. Up to now seven plants are in operation in Scandinavia. In addition to this the key component in the process, the separator, has been sold to a number of waste plants in Europe and Japan.

The RDF produced is fired in boilers, normally for hot water production feeding district heating systems, which are common in Scandinavia. A major advantage is that the RDF can be stored during the summer season.

ABOUT THE COMPANY

PLM Sellbergs is one of the major waste contractors in Scandinavia and quite sizeable also according to European standards. The company was founded in 1882. Turnover in 1987 was 470 M Sek (US $ 80 M) and we employ close to 1000 people in Sweden and almost as many in Spain.

Besides being involved in normal waste collection we also own and operate a number of large transfer stations and landfills.

Through extensive research in the 60´s and 70´s our engineering division developed a technology to separate household waste into RDF, compost & metal known as the BRINI Process. I will describe the process later but first a few words about the general situation for waste disposal in Sweden.

WASTE DISPOSAL IN SWEDEN

Household refuse in Sweden is disposed of in four typical ways: landfilling, massburning, separation and burning and composting. The

proportion are presently about 35 %
for landfilling, 55 % for massburning
and 10 % for the two last. What sing-
les out Sweden from most other countries is that energy recovery can be
quite easily obtained when burning
raw waste or RDF by feeding the hot
water from the incineration plants
into the cities' district heating networks. The successful application of
this technique has led local municipalities to talk about "energy production" rather than "waste disposal".
All larger Swedish cities - and many
smaller - already have extensive
district heating systems.

There are about 25 plants using waste
or RDF fuels. Some 20 are straightforward incineration plants. All of
them are feeding the produced heat
into district heating networks. Most
of these plants have a very good operating record giving satisfactory service to their customers.

The success of this combined waste
disposal/energy production technique
halted, however, when flue gas quality
requirements stiffened.

The existing mass burn plants were
built at a time when the flue gas cleaning requirements could be satisfied
with electrostatic precipitators,
sometimes with limestone injection
added.

In the last years Sweden has been
mostly known for its moratorium on
new mass burn plants. This moratorium
has now been lifted but with new environmental standards, which will
require a wet flue gas cleaning system
with mass burn plant.

These new standards have also resulted
in a discussion about presorting the
waste into RFD or similar.

WASTE AS AN ENERGY SOURCE

Let us take a look at waste as energy
source: It offers an even and stable
supply throughout the year, it is not
exposed to speculation and it is produced in reasonable proximity to the
user. It has several disadvantages,
though, one being that it is impossible to vary the supply to match the
demand.

The average ratio between summer and
winter needs in the district heating
is about 1:7, i.e. summer need is
approx. 15 % of winter need. In a
given network waste based energy can
be used up to 15 % of the maximum
need. In several cases large plants
receive more waste from surrounding
cities during winter season, leaving
these cities to find other means of
disposal during the warmer season.
The alternative, of course, is to
burn all waste throughout the year,
spilling out the energy without usage.

In the last years an increasing amount
of industrial process waste heat has
been introduced, in competition with
waste. Most district heating networks
have several heating stations using
different fuels, i.e. on marginal
cost calculation basis waste and other
solid fuels can be valued at oil replacement cost.

At present we produce 3 TWh from waste. Theoretically this could be expanded to 5 TWh. But the bigger cities
are now already done, so the possible
expansion is medium size towns, where
the scale factor makes mass burning
relatively more expensive.

WHY REFINE THE WASTE?

Why, then, refine the raw waste before incineration? There are some
obvious reasons:

Although it is quite possible to incinerate raw waste, it cannot be called a fuel. It has a highly heterogenerous composition, which may cause
the heat value to vary as much as
+ 50 % from the average value. Untreated waste also contains a lot of
materials with little or no energy
content.

A RDF gives an identifiable product

with stable heat value, lower ash content, simpler combustion, simpler flue gas cleaning, and it is more compatible with other fuels. Besides, it can be handled and stored easily, offering a way of adjusting the supply to the actual demand. The obvious drawback is of course the cost for the process, but in most cases this cost is offset by the cost for the boiler and the flue gas cleaning equipment.

When PLM Sellbergs set out to develop a technique for converting waste into fuel, we proceeded on the basis of market demands for solid fuel with a maximum energy content. The energy-bearing constituents in waste are mainly paper and plastic. We therefore concentrated on these materials, and developed a separator unit that could sort out mainly paper and light plastic into one fraction.

As untreated waste is a highly heterogenerous material, we used durable, heavy-duty equipment for handling and processing.

A waste-handling plant must also be highly efficient and highly reliable, as waste is generated every day and storage is more or less impossible.

The BRINI System meets all these requirements, and for this reason it is now the most generally used method for handling waste in Sweden. It should be noted that of the waste-to-energy plants built in Sweden in the 80´s almost half of them are using the BRINI-process.

THE BRINI PLANT AT KOVIK, STOCKHOLM

A total of seven plants for producing either RDF of dRDF with the BRINI System are now in operation in Scandinavia.

I will now describe the most advanced of these plants which is the Kovik plant near Stockholm, Sweden.

This plant comprises BRINI modules 1, 2 and 3. In module 1 waste is pulverized and separated before being pelletized (module 2). The screen fraction is used for production of compost (module 3).

This plant is owned and operated by PLM Sellbergs, Stockholm. It receives household waste and similar industrial waste from the municipalities of Nacka, Värmdö and Tyresö. Design capacity is 27 tons per hour. The plant was taken in operation in 1982.

The principal processing stages include receiving, pulverizing, separation, drying, pelletizing, cooling and discharging. (See Appendix 1).

Operation comprises the following stages:

Waste collected by compactor trucks is dumped either into a bunker or directly into the feeder. An overhead crane fitted with a grab transports the waste from the bunker to the plate feeder, which feeds it to the pulverizer, a Tollemache 1500. The pulverizer has a vertical axle and has no grates. It therefore consumes little energy.

The shredded waste is carried from the pulverizer on a belt conveyor to the flow divider and the flow distributors which feed the waste evenly to the three BRINI classifiers. Each classifier has a capacity of 10 metric tons per hour. The waste is mechanically separated into three fractions in the classifiers.

Fraction 1. The combustible material is concentrated in a light fraction, which consists mainly of paper and light plastic. Emission generators, metals and PVC are largely absent from this fraction.

Fraction 2. Other organic materials, glass and sand are concentrated in a screen fraction.

Fraction 3. The third, heavy fraction contains iron and other metals as well as various other residual materials, like rigid PVC bottles, etc.

Appendix 2 shows the material balance and appendix 3 the energy balance of the process.

The light fraction which is the input for Module 2 is then transported to a buffer storage unit which ensures uniform flow in subsequent stages. The moisture content of the waste is reduced to about 15 % in a dryer before it is fed into the two pelletizers, which convert it to pellets under high pressure. The finished fuel that emerges from the pelletizers is warm. It is cooled in a conveyor before being loaded into containers with capacities of 30 m^3 (40 yd^3).

The fuel output is 40-50 percentage by weight. The fuel produced has a uniform shape with a high effective heat value of 17 MJ/Kg (7300 Btu/lb), a moisture content below 15 % and an ash content of about 10 % (DS).

The pelletized fuel is marketed to external customers such as industrial firms and municipal district heating plants. Since November 1984 the whole production is delivered to the central heating plant of Drefvikens Energiverk AB (DEAB) in Bollmora 30 km from the Kovik plant. The plant is described further in the following section.

BOLLMORA SOLID FUEL BOILER

Until the spring 1984 Bollmora had a normal waste incinerator. Since 1969 household refuse had been fired on a one-shift basis. In 1984 the plant had to be closed because of bad conditions and also because of it being equipped only with a multi-cyclone for dust cleaning.

In March 1984 DEAB accepted the offer from PLm Sellbergs, as a turn-key contractor, to undertake a complete re-building of the old "waste incinerator" to the type "stayed water wall boiler". At the same time DEAB signed a long term contract for fuel supply from PLM Sellbergs, consisting of BRINI pellets (dRDF) and wood chips.

Today the Bollmora district heating central is equipped with a base load 10 MW boiler which is specially designed for the new fuel.

Principle arrangement of Bollmora solid fuel boiler is shown in Appendix 4.

The reconstruction mainly comprises:

- new fuel handling system
- new grate
- rebuilding of the boiler, a new convecting heating surface and an economizer
- new flue gas cleaning equipment
- new automatic clinker handling system
- new automatic control system

The plant is run in 3-shift operation and has now two shifts unmanned with so called periodic supervision.

Normally a fuel mix of approx. 75 % pellets and 25 % wood chips, calculated as energy equivalents, is fired. The mixture proportions can be variably adjusted in the metering bins.

A new electrostractic precipiator has been designed for clean gas with maximum 35 mg/Nm^3 dust content (at a CO_2 content of 10 %).

To be able to provide the district heating system with water with a temperature of up to 120°C (250°F) a completely new boiler circuit consisting of a so called series boiler (convecting heating surface) and an economizer has been added. The series boiler is hanging down like a cassette in the second zone of the boiler and is placed before the flue gas tube bank of the boiler. A so called economizer reduces the flue gas temperature to approx. $160^\circ C$ ($320^\circ F$).

The plant is now run in continuous 7-days operation. The plant had some starting-up problems but these have been solved gradually.

The greatest problem during the spring 1985 was clinker on the refractory lined walls. This was solved during the summer stoppage by installing air-cooled side walls. Another problem is clinker lumps in the grate, resulting in heavy wear of the ash screws. This problem has been reduced after the installation of a clinker crusher.

Today the plant is functioning well and produces all the heat the net needs in the interval 5-12 MW. The uniformity - in both size and composition - of the fuel makes it possible to control the combustion in a close interval with a low O_2-content of 5-6% and a CO-content of less than 100 ppm. The thermal efficiency of the boiler has been measured to be 88-90%.

SUMMARY

The BRINI process has proven itself in many applications throughout Scandinavia. We have also installed a number of separator units in plants in Europe and Japan.

The RDF produced is a quality fuel as compared to raw waste with high energy content and excellent handling and burning properties.

Finally I would add that we are represented in this country since about 4 years and are presently pursuing a number of interesting projects in the US and Canada. A contract has recently been signed for the first BRINI plant in Nanaimo, B.C., Canada.

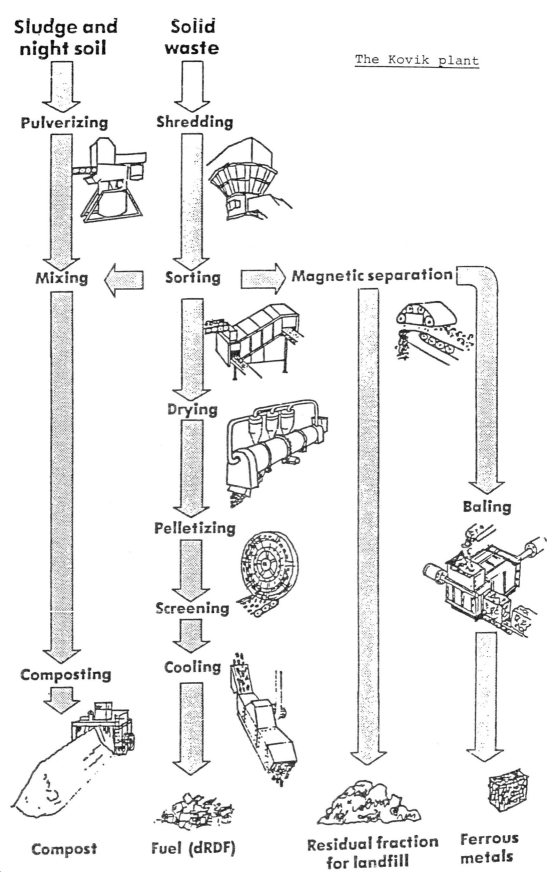

Figure 1.

Material balance

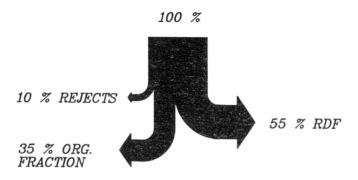

Figure 2. **PLM Miljöteknik AB**

Energy balance

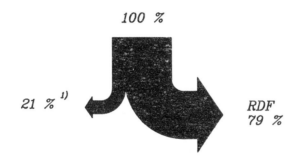

[1] About 10 % more energy can be recovered by unaerobic digestion of the organic fraction, giving the following balance.

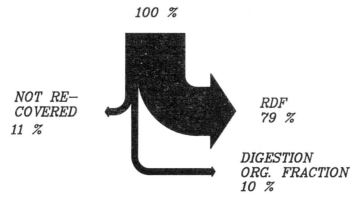

Figure 3. **PLM Miljöteknik AB**

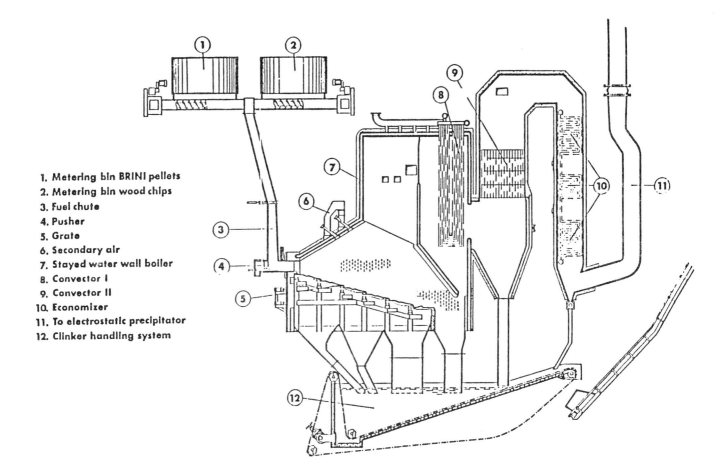

Figure 4.

THE VALORGA SOLID WASTE REDUCTION PROCESS FROM FRANCE

Lee J. Beetschen ■ CABE Associates, Inc. 144 S. Governors Ave., Dover, DE 19903
Philippe Cazanave ■ Valorga, Z.I. de Vendangues, 5, Rue de Nassacan, BP 56, F-34740, Vendangues, France

Valorga has developed an innovative process for the treatment of municipal solid waste (See Exhibit 1). After crushing and sample sorting, refuse is fed into an anaerobic digester with approximately fifteen (15) days detention time. It is important to note that mixing is accomplished without mechanical equipment. Under continuous methanization at high concentrations of organic matter (35-40% dry matter), the organic waste is transformed into biogas and a residual called "digestate". The biogas has a calorific value of 4,000 to 5,000 BTUs per pound and can be burned in a combustion unit to produce high and low temperature energy. The process generates between 3,500 and 5,000 cubic feet of gas per ton of refuse at a 60 to 65% methane content. The digestate is a high quality soil conditioner for agricultural and horticultural use, with 60% dry matter content and 40% of organic matter. The material is stabilized to the point where it can also be used for landfill cover. The first facility using the Valorga process started in 1984 near Grenoble. It has a capacity of 16,000 tons per year. A larger facility, 100,000 tons per year, is scheduled for start up in the city of Amiens, France, before June of 1988.

The continuous methane producing fermentation of high dry matter content substrates was developed as a result of a research program at French universities, with Valorga as an industrial partner. The first study phase concerned the laboratory development of pilot digesters with a capacity of 20 to 30 liters for continuous treatment of urban waste, with a dry matter content of between 30 and 35%. The purpose of these studies was to identify the critical design criteria for the operation. They were determined to be:

1. Fermentation Time

   This determines the degree of gasification and mineralization of the digestate. It, therefore, governs the biogas output per unit of matter and the agrinomic quality of the digestate.

2. Dry Matter Content

   This influences bacterial activity, with insufficient humidity being noted as a limiting factor in fermentation.

3. Concentration of Different Constituents in the Substrate

   In particular, the concentrations of carbonaceous and nitrogenous compounds which affect the density of the bacterial populations necessary for the digestion process.

4. The Quality of Mixing

This affects the ability of bacteria to obtain nutrients and the quality of thermal exchanges.

5. Regularity of Feeding

This factor influences the balance between the different populations, i.e. between the three (3) basic digestion phases.

In the second phase of process development, after using an experimental 108 cubic feet industrial digester, a treatment unit with a capacity of 18,000 tons per year was installed near Grenoble, France. It has been in continuous operation since then. In addition to the recently completed Aimens facility, four (4) other contracts have been signed for facilities in France, with capacities ranging from 20,000 to 100,000 tons per year. Before providing some historical perspective on the process, and a more technical description, it should be pointed out that the methanization unit has the capability of treating a variety of other waste if mixed with domestic solid waste, including sewage sludge, waste from the food industry, and animal manure.

The discovery of the methane fermentation process dates back to 1667, the year in which Alessandra Volta, the inventor of the electric battery, discovered the presence of methane in marsh gas. Since the second half of the 19th century, numerous experiments and actual full-scale facilities have utilized the anaerobic digestion process. Process development was in two (2) different directions, i.e. the batch and continuous process. The continous process has found particular success in sewage plants for the stabilization of by-product sludges from the primary and secondary unit treatment systems. Until Valorga developed the system at Grenoble, the process had been mainly relegated to low solids content systems. It is for this reason, that this paper concentrates on a discussion of the methanization process. The preliminary crushing and sorting steps and the final refining steps are not unique to the Valorga system. The methanization step can be integrated with existing systems and, in many cases, proves to be most cost effective in such a setting.

The mixed bacteria population initially breaks down biopolymers that make organic matter into substances that microogansisms can more easily metabolize. This initial phase, which is essential to the success of fermentation, is more difficult to achieve with a ligno-cellulose substrata, such as manure, because through incrustation and impermeabilization of the cellular components, the lignin impedes the hydrolysis of other components. Acidogenic bacteria transforms the simple compounds obtained after hydrolysis into a complex mixture rich in volatile fatty acids. They also transform the compounds into alcohols and other molecules having more than two (2) carbon atoms, and into carbon dioxide and hydrogen. The methanigenic bacteria use only 10% of these constituents to create additional bacteria. The remaining 90% are transformed into methane and other gases, carbon dioxide and hydrogen sulfide in particular. Researchers in the early 1980's isolated an intermediate stage, called "acetogenesis". This process transforms the organic molecules having more than two (2) carbon atoms into acetic acid, carbon dioxide, and hydrogen, the precursors of the final gas mixture called "biogas". Methanigenic bacteria cannot directly assimilate these complex organic molecules.

When the dry matter content of the substrate to be methanized is less than 10%, the fermentation can almost always take place in a continuous process. The very diverse continuous processes - completely mixed, recycled-sludge, sludge-blanket, and packed-bed processes using anaerobic filters or fluidized beds - all have two (2) advantages as compared with batch processes; A single digestion chamber yields steady biogas production, and the loading and unloading oeprations are amenable to automation.

Though not new, the idea of implementing continuous methanization of solid waste having a high dry matter content suffered from the difficult problem of how to circulate an irregular and unwieldy substrate through the digestion system. Pumps proved to be ineffective or to consume excessive amounts of energy; moreover, the inescapable wear and tear on components made it impractical to use conventional mechanical systems. Nonetheless, the intriguing fact that biogas is being recovered from domestic landfills in the USA, Germany, the United Kingdom, and France, provided obvious proof that domestic solid waste is an eminently suitable substrate for methanization. Indeed, the methanization concept remains valid even though, because of evolving lifestyles, the waste contains increasing amounts of non-biodegradable material at the expense of organic matter that bacterial can assimilate.

Valorga solved the loading and mixing problems. A piston pump loads substrate into the digester inlet on a continuous basis, and at the same time, a similar quantity of material is extracted from the outlet. A pneumatic system having no internal mechanical parts transfers the substrate into and out of the digestion system by means of a set of biogas-valves and chambers at the inlet and outlet. Mixing takes place through the programmed injection of pressurized biogas into the system's different sections, depending on the different stages of methanization.

The Valorga process excels by its adaptability to the methanization of polysubstrates, its ability to methanize a high dry matter content substrate as high as 40%, yielding a digested effluent of about 60% dry matter, the continuous loading feature of the digestion system and, finally, the pneumatic mixing.

The operation is extremely efficient. Whereas conventional processes produce approximately 1 cubic meter of biogas per cubic meter of tank volume per day the Grenoble system produces 5 cubic meters per cubic meter of tank volume. Further, only 20% of the raw material has to be rejected or landfilled. All the other products produced are saleable, albeit market dependent.

A 100,000 ton municipal solid waste facility, such as the one currently undergoing start up at Amiens, provides a good overview of the process. It is estimated that approximately, 1,000 tons will be in the form of uncrushable material that has to be landfilled. Ferrous metal will constitute about 3,800 tons and glass 6,200 tons. Before water is added for conditioning the material for introduction into the digester, another 20,000 tons of combustible refuse derived fuel is removed. Recycled water is added to the remaining mixture which is then fed into the digester, which produces 12,000 tons (320 million standard cubic feet) of biogas and 1,000 tons of water which escapes with the gas. After pressing, 49,000 tons is delivered to a unit which crushes the dry digestate and produces 13,000 tons of combustible reject and 36,000 tons of the agrinomically important digestate. To close the material balance, 7,000 tons pressed material must be landfilled.

From a physical standpoint, the system is relatively compact, requiring only four and one-half (4 1/2) acres for a 100,000 ton per year unit. These acreage requirements are:

Buildings	45,000 sq. ft.
Digesters	10,000 sq. ft.
Refined Digestate Storage	100,000 sq. ft.
Utilities and Green Spaces	40,000 sq. ft.

The process consumes about 800,000 gallons of water, 3,100 WMh of electricity, and heat requirements are 4,100 MWh.

In conclusion, Valorga offers some rather unique qualities:

1. It can be integrated with any number of existing solid waste handling facilities which already have crushing and sorting operations to remove glass, ferrous, and non-ferrous metals.

2. The methanization process provides two (2) saleable products, biogas and digestate.

3. Prior to being able to market the digestate, the product because of its stability, can be used for landfill cover.

4. If placed in proximity to an existing landfill undergoing gas recovery, in the short term it can add to the volume of gas produced and in the long term provide a steady source of biogas to the consumer.

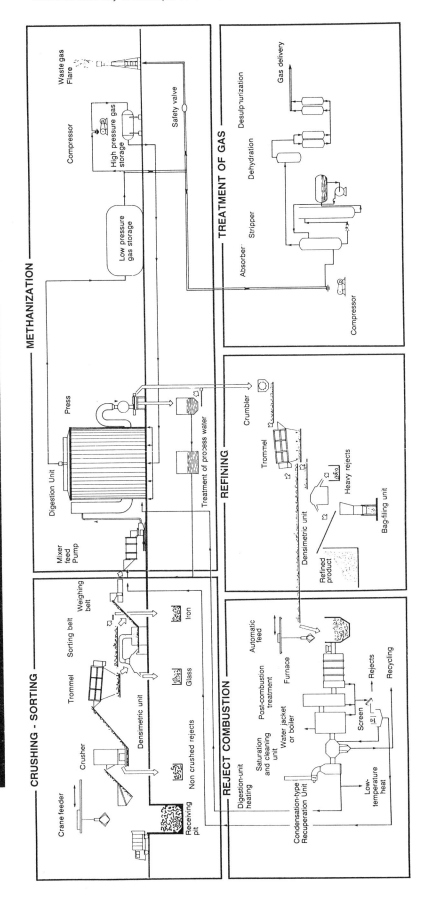

Exhibit 1.

WASTE MANAGEMENT IN HOLLAND AND THE LARGEST INTEGRATED WASTE PROCESSING PLANT IN THE WORLD

Arian Bos ■ N.V. Akvalverking Rijnmond, P.O. Box 1120, 3197 KK Rotterdam, 3180 AC Rozenburg, The Netherlands

I would like to tell you something about the waste policy in the Netherlands and I will explain that by experiences and developments at AVR (Afvalverwerking Rijnmond).

The Netherlands is only a small country; with an area of 36600 square kilometers it is eleven times smaller than your State of California.
Most Americans know the Netherlands (or Holland as it is often called) only in connection with windmills, tulipfields and wooden shoes. It may come to you as a surprise that also the largest integrated waste processing plant in the world, with a processing capacity of more than a million tons of waste per year, is situated in that same small country.

This does not mean that its inhabitants, the Dutchmen, are such tremendous waste producers; the 340 kg of domestic waste produced by the average Dutchman per year is considerably less than the corresponding American figure. The reason for this difference is to be found in the high populationdensity and the geological and climatological conditions there.
I already made a comparison with the State of California; the average populationdensity in the Netherlands is eight times higher. This means that we cannot afford to plainly dump our waste, we simply don't have enough space available to do so. As a consequence of the high groundwaterlevel (more than one third of the area of the Netherlands lies below sea-level) and a yearly rainsurplus of 350 liters per square meter, very stringent measures have to be taken against leaching out, and dissipation into the groundwater of toxic compounds and heavy metals, when installing landfilling sites.

The waste policy of the Dutch government sets the following order of priorities:
1. Limitation of wastegeneration
2. Recycling and re-use of waste materials
3. Incineration with recuperation of energy
4. Landfilling

A. Bos is production manager of Afvalverwerking Rijnmond, Rotterdam Holland

Despite the requirement mentioned under 1. the production of domestic waste in the Netherlands has, after a few years of moderate growth, suddenly increased with 10% in 1987 compared to 1986.

The requirement mentioned under 2. is more succesful. Last year 53% of used paper was collected separately in the households and almost half of the number of circulating glassbottles was collected in the waste-glass containers specificly placed for that purpose at shoppingcenters. Also the separate collection of garden- and kitchenwaste for compost production slowly gets going. To put an end to the simple dumping of refuse the Department of Housing, Geographical Allocation and Environment, has set up a plan that visualises the construction of five to six large new waste-incineration plants in the next five to ten years.

These installations will have a processing capacity of 350.000 to 600.000 tons a year and will be equipped with steamboilers and turbine-generators for recuperation of energy. These new installations will replace old and small installations and landfilling sites still in use and at the same time cope with the growth of the waste production.

Stringent emission-limits will be set for these new installations:

	24-hours average	maximum T.L.V.
HCL :	50 mg/Nm3	75 mg/Nm3
HF :	3 mg/Nm3	5 mg/Nm3
Dust:	50 mg/Nm3	75 mg/Nm3
Pb+Zn:	5 mg/Nm3	
Cd :	0,1 mg/Nm3	
Hg :	0,1 mg/Nm3	

The expression Nm3 refers to a recalculation of the actual value to a normalised state of dry fluegas with 11% O2 at 101.3 kPa.

By setting a limit for the emission of dust consequently also the emission of contaminants present on or in the flyash particles, like heavy metals or organic trace-contaminants (PCB, dioxines) is controlled. Besides, along with these abatements of consequences also a lot is being done in the field of prevention at the source. Cadmium, for instance, shall only be used for those applications for which no alternatives are available yet, and the mercury content in dry batteries shall be lowered by a factor eight between now and the end of the year 1990. After these general discussion of the official waste policy in the Netherlands, I would like to tell you something about the company I am working for: AVR.

Its plant, located in the Rotterdam industrial- and harbour area was put into operation in 1972 and has remained since then the world's largest waste processing plant. It has six moving grid furnaces for the incineration of domestic- and industrial waste and two rotating drum furnaces for the incineration of chemical waste. The grid furnaces, supplied by the "Deutsche Babcock" (represented in the USA by its affiliate: Ford, Bacon and Davis) each have seven rotating grids. Their incineration capacity of 21,5 tons per hour at a grid width of 4,20 metres is high, compared with that of reverse-acting grids. With the combustion of the waste as delivered to AVR with a heating value of 8,000 kJ/kg, out of every ton of waste 2,0 tons of steam are produced. This steam has a pressure of 27 bar (2,700 kPa) and a temperature of 370 degrees Centigrade (700 degrees Fahrenheit). This pressure and temperature are both relatively low. In modern installations it is quite usual to operate with a pressure of 40 bar (4,000 kPa) and a temperature of 400 degrees Centigrade (750 degrees Fahrenheit).

Limitative in this respect is that the pipe-wall temperature under no circumstances shall exceed 550 degrees centigrade (1020 degrees Fahrenheit) to prevent high-temperature HCl-corrosion.

The grid furnaces have two-draughts boilers, with a high (22 metres) empty first draught. This part, consisting of tube walls (membrane walls) receives mainly radiation heat. The flue gasses cool down to such an extent that no sticky fly-ash particles are present that could cause premature plugging of the boiler. In the second draught there are in sequence, boiler tubes, superheater tubes and the economiser. The steam is fed partly to a drain/condensation turbine of 28 MW, where with every ton of steam about 200 kWh of electric energy is generated. Another part is fed to two backpressure turbines of 14 MW each. Here the steampressure drops to 2.5 bar (250 kPa). This lowpressure steam is fed to a waterdistillation unit. Here in three multi-stage flashevaporators with every ton of low-pressure steam about nine m3 of distilled water are produced. This distilled water is transported to the Rotterdam industrial area by means of a special pipeline system. It is used as boiler feedwater and proceswater by the chemical and petrochemical plants there.

In 1987 954.000 tons of waste were processed in the grid furnaces whereby 240.000 MWh of electric power and more than 6 million tons of distilled water were produced. Besides these useful products fly-ash and crude bottom-ash are made (3% and 26% by weight of the processed waste respectively). The fly-ash is partly used as an additive in bitumen. The remainder is very unfortunately still deposited on a controlled filling site, specificly designed and equipped for that purpose.
The crude bottom-ash is converted entirely into useful products in the bottom-ash processing installation which was put into operation in 1986. With this AVR contributes to the Dutch goal to burden the landfilling sites as little as possible.
More about this installation will be told later in this lecture.

Two thirds of the domestic waste is delivered to the plant by truck (directly by the waste collecting truck or by bulk carrier trucks from re-load transfer stations). The amount of 300.000 tons per year coming from the province of Utrecht is transported in presscontainers by barge over a distance of 96 km (60 miles).

Now I would like to tell you something about the two rotating drum furnaces. These are used for the processing of chemical waste and that part of the waste coming from hospitals that is not allowed to be processed in the domestic waste processing furnaces.

The chemical waste can be processed as a liquid via a tankfarm, as pasteous material via a waste storagepit, or packed in metal or plastic containers of drums.
In the year 1972 two rotating drum furnaces were put into operation, without boilers and with an elctrostatic filter as the only means of fluegas purification. Meanwhile one of these furnaces has been replaced by a completely new installation provided with a boiler, electrostatic filter and wet fluegas purification. This installation, at the moment the most modern one in Europe, has a capacity of 45.000 tons per year. This drum furnace has been operational for a year now, and has recently gone through a PCB combustion test in which the severe demand of 99.999% conversion and breakdown of these toxic components was fully met. The furnace with all its appendices and additional provisions was constructed by the Swiss company Widmer + Ernst (represented in the USA by its affiliate Combustion Engineering).

I already promised to say something more about the new crude bottom-ash processing installation of AVR, where only the bottomash of the domestic- and industrial waste incineration is being processed.
This installation was set up to convert the crude bottom-ash, which constitutes 26% by weight of the amount of waste processed, entirely into useful products. The installation was further developed from a basic design by AVR, in co-operation with the German company Lindemann (active in the USA under the same name).
I already mentioned that from 1.000 kg of waste, besides the 2.000 kg of steam, also 30 kg of fly-ash and 260 kg of crude bottom-ash remain. Our bottom-ash processing installation separates this crude materials into:
- 223 kg of clinker (with particle size 0-40 mm)
- 21 kg of scrap steel
- 7 kg of non-ferrouw mix
- 9 kg of incompletely incinerated material

These data refer to average values over a period of a year.

The clinker is used as a base foundation layer under roads, etc. and for construction of dikes, noise-abatement- and wind-screens. In connection with the specific Dutch circumstances mentioned earlier, like high groundwaterlevel and rainsurplus, the use of this clinker is legally restricted to covered applications only, and at least 50 cm above the highest expected groundwaterlevel, to prevent leaching of heavy metals.

The scrapsteel has a Fe-content of more than 93% by weight and a density of 1.2-1.3 kg/dm3 and is suitable for use as a raw material in the steel industry.

The non-ferrous mix consists of a mixture of mainly copper, aluminium and brick. This mixture is supplied to a company that recovers the pure metals by means of Eddy-current- and flotation techniques.

The incompletely incinerated material is transported back to the waste storage pit of the waste incineration furnaces.
The lay-out of the installation can be seen on this picture.

Via the conveyor belts (1) and the swing conveyor the crude bottom-ash from the waste furnaces is dumped on a concrete floor in a halfcircular pile (2). This supply continues round the clock. The storage capacity of the floor has been designed to store the crude bottom-ash resulting from five days full-shift operation of the waste furnaces. In daylight operation the bottom-ash first stored (so consequently optimally dewatered) is put on a steel feederbelt by means of a shovel.
Then, the bottom-ash goes via the conveyorbelt (3) to the firms drumsieve (4). This drumsieve has holes of 40 mm diameter. The overhead magnet 5 takes out all the steel components. This fine scrapsteel fraction (C) is supplied to steelmills to be used as an additive to the ore input. The clinker goes to the storage bin (B). The overflow, larger than 40 mm diameter, goes via a steel feederbelt to the hammermill (8). There is also a possibility to feed large metal objects like refrigerators, washing machines and bicycles that have been separated from the domestic waste. This hammermill has been subject of an intensive search and a great deal of extensive testing with our specific material. Out of all the various types and brands on the market we finally chose the Zerdirator of the company Lindemann. This equipment is well-known in Europe for scrap steel processing.
This hammermill performs a number of operations:
metal objects are hammered into clean, compact lumps about fist size large, bricks are granulated and the incompletely incinerated material, if present, is ground to suitable dimensions for an additional combustion-cycle; combined metals in appliances (like for instance the copperwindings in a steel housing of electro motors) are completely separated from each other by the hammermill. The hammered material first passes a drum magnet (9) where the steel components are taken out. The remaining material goes to the second drumsieve (12). This drumsieve has holes with an effective diameter of 12 mm. The ground bricks pass through the holes and are added to the clinker in the storage bin (B). The non-ferrous metals and the incompletely incinerated material are fed into a air-classifier (13). The air-classifier produces a non-ferrous mix as its bottom product. The incompletely incinerated material goes over the top and gets via covered conveyorbelts in the presscontainer (15), in which it is transported back to the storage pit of the waste furnaces.

80 Resource Recovery of Municipal Solid Wastes AIChE SYMPOSIUM SERIES

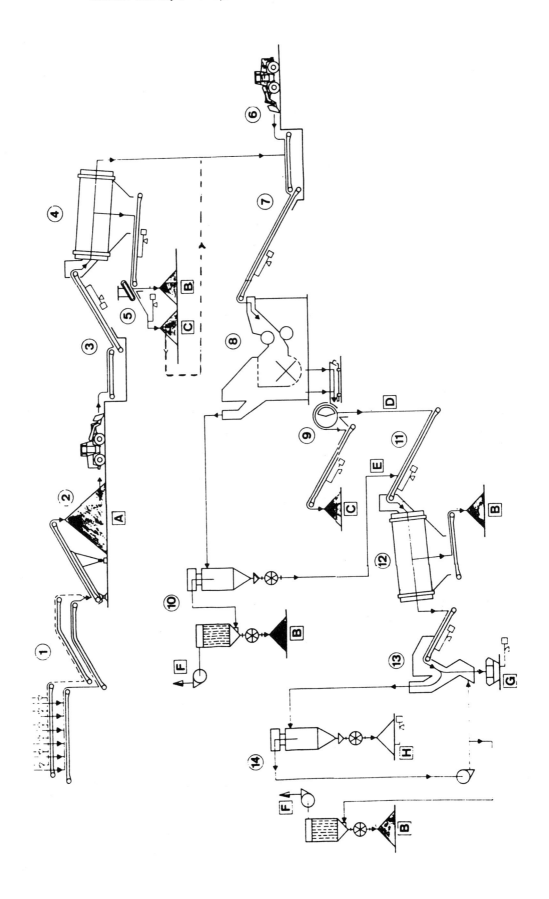

—AVR Rotterdam—

THERMAL AND BIOLOGICAL OPTIONS FOR ADVANCED RDF SYSTEMS

Donald K. Walter ■ Biofuels and Municipal Waste Technology Division, U.S. Department of Energy, Washington, DC
Barbara J. Goodman ■ Energy from Municipal Waste Program, Solar Energy Research Inst., 1617 Cole Blvd., Golden, CO 80401
Christina E. Thomas ■ Solar Energy Research Institute, 1617 Cole Blvd. Golden, CO 80401

Municipal solid waste generated by all sectors of the economy could potentially be converted to energy. This paper describes the mechanical processing necessary to produce refuse-derived fuel, the feedstock that can be converted to steam, electricity, and liquid and gaseous fuels. Also described are the thermal and biological processes that lead to production of renewable liquid and gaseous fuels. Current commercial status of various systems and future research direction are also described.

Municipal solid waste (MSW) is generated by the residential, institutional, and industrial sectors of the economy. Each year approximately 200 million tons of MSW are generated in the United States. Waste by definition has negative connotations, but it can be used to generate energy; the energy potential is estimated to be equivalent to 300-400 million barrels of oil per year.

MSW is heterogeneous, and even the quantities of any particular constituent vary widely from one location to another. This characteristic means that MSW presents a formidable barrier to the effectiveness of unit operations in processing systems designed for more heterogenous materials. Table 1 shows the average constituents compiled from a series of Department of Energy reports. Mechanical processing systems produce the solid refuse-derived fuel (RDF) that is used directly as a boiler fuel or as a feedstock for thermochemical and biochemical conversion. Figure 1 shows the options for converting MSW into different forms of energy.

Thermochemical conversion produces steam, electricity, or liquid and gaseous fuels from RDF by the application of heat. The oldest thermal technology for producing energy from waste is combustion. However, gasification and pyrolysis systems are emerging. Pyrolysis and gasification systems produce liquids, gases, and solids from MSW. Pyrolysis is the chemical degradation of a substance by heat, usually in the absence of oxygen, while gasification processes chemically degrade substances with restricted oxygen.

Table 1. Constituents of Municipal Solid Waste

	Constituent Weight Percent	Heating Value (KJ/kg)
Paper and paper products	37.8	1750
Plastic	4.6	3350
Rubber and leather	2.2	2350
Textiles	3.3	3250
Wood	3.0	2000
Food wastes	14.2	1510
Yard wastes	14.6	1700
Glass and ceramics	9.00	
Metals	8.20	
Miscellaneous	3.10	
TOTAL	100.0	

The moisture content averages about 25%; the inorganics average about 20%; the ash in the organics averages about 5%.

Compiled from a series of Department of Energy reports published between 1975 and 1982.

Biochemical conversion uses microorganisms or enzymes to convert MSW to liquid or gaseous fuels. Anaerobic digestion is the process in which microorganisms convert the natural organic matter in MSW to a mixture of methane and carbon dioxide. Certain enzymes selectively convert cellulose to glucose, which may then be fermented to produce alcohol.

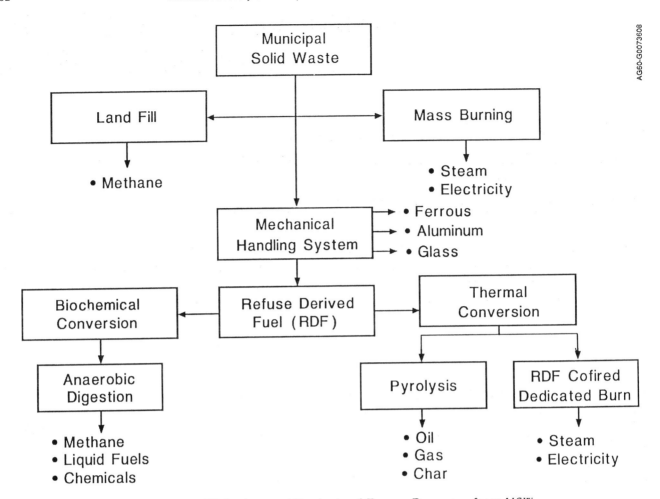

Figure 1. RDF Options and Products of Energy Recovery from MSW

MECHANICAL SYSTEMS

Mechanical systems that separate waste into specific components to be recovered and used again represent a young technology. These systems perform size reduction, separation, and RDF optimization. The size-reduction equipment used in plants that mechanically process MSW was developed for other feedstocks and must be modified for use with MSW, where its performance is often less than optimum. Reducing MSW to a smaller size also produces more uniform particles for subsequent separation operations.

Hammermills reduce waste size through a combination of impact and shear. In a horizontal mill, the waste is forced through a grate. Residence time in the mill and particle size produced are largely functions of the size of the openings in the grate; however, moisture content, feed rate, waste composition, hammer spacing, and rotor speed also play a part. A vertical mill does not have grates, and the particle size produced depends principally on hammer spacing and wall clearance. In general, hammermills use a great deal of energy, with 750-kW (1000-hp) motors. They are also high-capacity machines (with primary shredders in the 50-ton/h range) because of the size of the opening required to process MSW.

A flail mill also has hammers attached to a horizontal shaft, but it has no grate and may have one or more shafts. The size distribution of the product depends on rotor speed, hammer spacing, and hammer clearance. The flail mill consumes less power than the hammermill and generally suffers less wear because hard-to-shred items (such as metal bars) tend to fall through the machine.

Another type of machine used for size reduction is the slow-speed shear shredder. This device has two horizontal shafts with notched intermeshing cams. Waste is torn and sheared into pieces. Whenever a material that cannot be sheared is encountered, the machine automatically reverses and clears itself. Compared with other shredders, this machine is simple, durable, and a low energy consumer. Perhaps most important, it does not embed glass shards in paper fiber; this advantage permits the production of lower-ash, less abrasive fuel.

The wet pulper, developed for use in the paper industry, resembles a household blender, with a vertical shaft and blades turning in a water/MSW pool. The paper and food waste components become individual fibers; glass, cans, dirt, and gravel break up and pass through 2-cm (3/4-in.) openings in the bottom of the pulper. Larger materials that cannot be reduced in size are expelled in a junk chute. The wet pulper produces the best materials from raw MSW; however, it is more expensive to own and operate than other size reduction equipment.

Reduced MSW can be separated into a number of components. The nature of the separation equipment depends on the physical and chemical properties of the material to be recovered. If the prime interest is in the production of fuel, then noncombustible matter must be removed. The most developed recovery equipment uses magnets to recover iron, steel, and other ferromagnetic metals. Magnets work well and are highly reliable. The principal problem with their use is the difficulty of freeing the recovered metal from other elements.

Screens separate waste based on its physical dimensions. The desired characteristics of screens--beyond the size of the opening--include the ability to agitate the shredded waste to expose fresh material to the screen and the ability to clear material that blocks openings. The three types of screens currently used in mechanical processing plants are flat, trommel or rotary, and disk screens.

Flat screens use vibration to move material along the screen surface and to dislodge undersized particles trapped on top of large items. Although they are not often used in MSW plants, they have been used to separate a stream from a preceding separator.

The most-used screen to date, the trommel or rotating screen, is a rotating barrel with perforated sides. Frequently, trommels are built with screens of different sizes along their length so that three or more sizes may be recovered in a single device. Using trommels to open bags, break glass, and remove undersized material before it reaches a primary shredder has been particularly effective. Trommels are subject to jams from long stringy material and wire trapped in the holes.

The disk screen consists of a series of intermeshed disks, cams, or star wheels mounted on cylindrical shafts. Material bounces as the wheels rotate, and the undersized material falls through openings between the disks and their shafts. These devices are less susceptible to jamming than are trommels.

Air classifiers separate materials by subjecting the waste to an air current; the current drops heavy materials according to their size, density, and aerodynamic properties. The horizontal air classifier (air knife) is usually ternary, with three compartments to catch the extracts. The vertical, zigzag, and inclined air classifiers are binary devices that separate fractions by weight. The vertical air classifier has straight vertical sides and a vertical stream of air. The zigzag classifier is also vertical; however, the sides are accordion shaped. The inclined classifier has two air streams--one rising from the bottom through a bed of material and the second sweeping across the top of the bed to pick up the light fraction. The inclined rotating-drum classifier operates on a stream of raw MSW. The rotating drum breaks open bags and tied bundles of paper, and suction at the upper end of the drum removes lighter materials. The drum is perforated at its discharge end to function as a trommel. The claims for this device include a dramatic reduction in the incidence of explosions in downstream shredders. In a pulsed air classifier being developed at Duke University, uniformly intermittent pulses of air are applied to a vertical stream of shredded waste to retain the separation within the boundaries governed by particle density.

Numerous other separation techniques have been tried, with varying success. Glass has been successfully separated from sand and gravel by froth flotation, a wet process that selectively attaches air bubbles to the material to be recovered. Another wet process, heavy-media separation, depends on the characteristic densities of various target materials; it primarily separates various inorganic fractions. The heavy-media separator contains one or more liquids of varying densities. For example, one fluid might float aluminum while a second might float steel.

The principal product of the mechanical processing system is RDF. In its original shredded and separated form, it is fluff RDF. This material has a low bulk density and a relatively high moisture content, so it is difficult to store and transport. There are cases on record where as much as 2000 tons of shredded waste stored for a week solidified into a paper maché-like mass. To improve the characteristics of fluff RDF, research was undertaken to adapt the various pelletizers, cubetters, and briquetters that had been developed for the agricultural and charcoal industries. They either force shredded and separated waste through a die (pelletizer and cubetter) or squeeze it between two wheels. All function reasonably well, although in the United States the devices do not reach their rated capacities. The United Kingdom has been very successful in producing densified RDF for their industrial boilers. (The United Kingdom produces 70% of its steam for industry in coal-fired boilers.) Another effort to improve the characteristics of RDF involves treating it with acid. The acid embrittles the material, which is then ground in a heated ball mill. The resultant powdered RDF is very uniform and quite dry. Unfortunately, the first full-scale plant encountered many difficulties, and the developer went bankrupt before these could be resolved and the plant made commercially operational. On a positive note, however, the fuel produced by this plant was burned successfully in a cyclone utility furnace at up to 50% powdered RDF by Btu content.

In a mechanical processing system, the size reduction, separation, and RDF optimization components are connected by conveyers. The most prevalent conveyer is a moving belt or vibrating floor. These conveyers generally have served well, although points of intersection (particularly those making angled turns) have been subject to jamming. A second type of conveyer uses air to transport finished RDF from the processing plant to the storage bin or furnace. The principal problems have been in elbow wear and the detraining RDF from the air. Some systems have had problems keeping the RDF in suspension. A third type of conveying system uses pipes and pumps to convey wet slurries in plants using wet processes. No tool has been available to assess the effect of one unit process on other unit processes. However, under both federal and private- and public-sector sponsorship, a series of computer models has been prepared and partially verified. These models have been assembled into a master model named GARB, available for use on the IBM compatible series of personal computers.

The mechanical processing systems built to date were designed for recovering specific materials for recycling, for recovering RDF to be used in an existing boiler, and for producing fuel for a special boiler designed for the RDF. In the United States, the original intent of mechanical processing was to provide fuel to large suspension, coal-fired utility boilers. Of the 22 mechanical processing plants built in the United States, only two have been designed by the same engineer for the same purpose (i.e., as integrated, RDF-dedicated boiler plants), and even they are not identical; one is twice the size of the other and it accepts coal as an alternate feedstock. They were also designed at the same time so that neither could gain experience from the other. The remaining 20 plants are different designs but were also designed at the same time, and again, their designers were unable to draw upon others' experience. This provides some explanation for the relatively poor experience with these systems in the United States. Of the 22 U.S. plants, two were designed to recover materials, and both are closed. Neither plant was able to develop acceptable markets for the materials they produced, and both had technical problems that were not fully resolved. Eleven plants were constructed principally to recover RDF and secondarily to recover materials. Four are operational and sell RDF to existing utility boilers. The plants that are closed had a combination of technical and institutional problems. Nine plants have been built to provide fuel to a dedicated boiler. Six are operational, and the boiler for a seventh is in construction.

In the United Kingdom, a slower, more orderly research program developed. The English believe that the collection of sufficient waste in one place to make a significant impact on the fuel consumption of a large utility boiler would not be economical in their situation. Also, 70% of English industry uses small coal-fired boilers with grates. They have developed systems to fit these boilers. Their early work developed fluff RDF as feed for these boilers plus cement kilns. After success with these efforts, they began developing pelletizers to extend the utility and enhance the fuel value of RDF. Currently they have two RDF plants in operation and two densified RDF pilot plants gathering data.

Research and full-scale test facilities are also operating in Europe and Japan. These units and tests are meeting with varying degrees of success; most are reporting some difficulty.

THERMOCHEMICAL SYSTEMS

In the late nineteenth century the first incinerator was built to reduce waste volume, and by 1905 the logical step of adding a boiler to produce steam was taken in New York City and in Hamburg, Germany. Combustion technology continued to develop in Europe, where energy costs were high, land was scarce, environmental concerns emerged somewhat earlier than in the United States, and municipal district-heating systems provided an excellent market for the steam generated. Refractory-walled incinerators with waste heat boilers simplified pollution control and recovered energy, and the more efficient integrated water-walled boiler was built in the 1950s.

The United States, with cheaper energy and more available land, relied primarily on landfills to dispose of waste. On the East Coast, incinerators were used, but they lacked proper pollution-control devices and were overloaded. The environmental legislation of the 1960s, with its requirements for air-pollution control, forced the closing of most U.S. incinerators. Solid-waste legislation of the 1970s led to experimentation with pyrolysis and use of RDF in the United States. Europe and Japan continued to use boiler systems, with greater emphasis on energy production. In the 1980s, the United States seems to have rediscovered the combustion system, and Europe and Japan are developing RDF and pyrolysis systems.

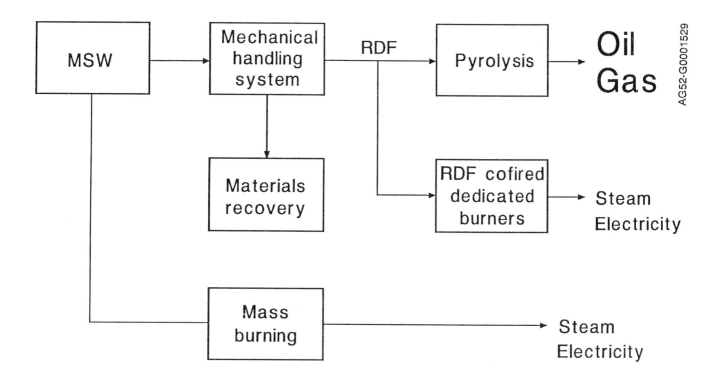

Figure 2. Thermochemical Conversion of MSW

Thermochemical systems can change RDF or raw MSW into such energy forms as hot water, steam, electricity, or valuable fuels (see Figure 2). More than 500 systems worldwide are burning waste to produce steam, but there are as yet few commercial-scale systems producing fuels from waste.

The distinguishing characteristic among thermal systems is the amount of oxygen made available. If sufficient oxygen is available to complete the oxidation of the components, then the system is an incinerator. If the oxygen supply is restricted, the oxidation is less complete; the resulting products comprise a range of combustible gases, liquids, and solids. If the heat produced by an incinerator is used for a productive purpose, then the system is defined as a combustor or boiler.

If the available oxygen is limited as MSW is heated, then the combustion process is not completed. The gases, liquids, and solids produced are fuels that can be stored, transported, and used for more purposes than the steam or hot water from a combustion process. There are many ways to describe these oxygen-limited systems. One major variable is the method of heating the feedstock. Direct heating methods burn a part of the feedstock in the reactor. If air is used, the nitrogen and other components limit the gas produced to an energy content of 5.6 MJ/m^3 (150 Btu/ft^3); if oxygen is used, the gas produced may have an energy content as high as 15 MJ/m^3 (400 Btu/ft^3). Indirect heating involves the application of heat through some external means. Since no oxygen is supplied to the reactor and there are no internal combustion products, the resulting gas typically has an energy content of 18.6 MJ/m^3 (500 Btu/ft^3).

A second means of categorizing systems is by their end products. All systems produce liquids, gases, and solids. The direct-heat systems use the solid product as a fuel and are generally designed to maximize the gas product. The indirect-heat systems are designed to maximize two of the three product forms and use the third as an energy source. The actual solid product is a combination of carbon char and ash. The gaseous product is a mixture of carbon dioxide, carbon monoxide, and hydrogen (with lesser quantities of higher hydrocarbons such as methane, ethane, etc.). The liquid products are highly complex hydrocarbons.

The third means of categorization is by reactor type. The variations are almost endless. The units may be vertical or horizontal. The gas may travel in either an updraft or downdraft direction. The bed of material may be solid feedstock or an inert material (e.g., fluidized bed).

All combustion systems have the following components: receiving area, storage area, preparation system, feed system, reaction chambers, energy or product recovery systems (boiler), pollution control equipment, and ash removal equipment.

The receiving area is where trucks dump their loads. If the plant is a modular combustor, or if it includes a mechanical processing system to prepare RDF for a dedicated boiler, then the floor is also a storage area. The storage area smooths the transition between the collection system, which typically operates 8 hours a day, 5-6 days per week) and the energy system, which typically operates 16-24 hours a day, 7 days per week. Large-scale combustion systems generally have storage pits, and RDF and similarly scaled systems have storage floors. The preparation area may be as complex as a mechanical processing system described above or as simple as a crane bucket mixing the waste in the pit. The feeding system depends on the design of the combustor. Most large mass-burning systems use a crane and bucket to transfer waste to a feed hopper. The RDF-dedicated boiler usually is fed by a series of conveyors and an air-swept spreader-stoker. The metering device can be variable-speed screw or belt conveyors, simple flappers, or some other device.

The combustion chamber, the heart of the system, provides a surface on which the waste can burn. The typical small combustor has a heat-resistant surface and hydraulic rams to move the burning waste through the furnace, while the mass-burning furnace has a grate that supports and transports waste and admits air under the burning refuse. The RDF-dedicated boiler usually has a traveling grate to perform the same functions. The walls may be constructed of refractory materials (refractory wall) or of welded tubes through which water circulates (water wall). In the fluidized-bed combustion system, the chamber is a box of sand made fluid by a rising current of air. The hot sand bed provides a thermal mass to ensure rapid and complete combustion. The fluid bed chamber can be constructed with water walls.

The typical boiler is a series of tubes hung amid the hot combustion products from the burning of the waste. In very small installations, the hot gases might circulate inside the tubes (firetube boiler). The arrangement of the boiler tube banks varies with the purpose of the installation. The boiler may have superheater tubes, an economizer to assist the water walls in converting boiler water to steam, feedwater heater tubes, and combustion air preheaters.

The pollution-control equipment cleans the stack gases before they are discharged to the atmosphere and monitors for potential problems such as dioxins and furans. Electrostatic precipitators or fabric-filled baghouses remove particulate matter. Wet and dry scrubbers remove sulfur oxides and hydrogen chlorides; however, in the United States, HCl is not a controlled pollutant as it is in many European countries, and MSW is a low-sulfur fuel. Overall acid gas emissions from an MSW combustion system average about 300 to 400 ppm, which compares favorable to the 800 ppm for low sulfur coal or oil and 2500 to 4000 ppm for high-sulfur coal or oil per unit of fuel.

Ash-removal systems depend on the combustion-chamber design and the type of ash. Siftings are the ash that falls through the grate; bottom ash is the unburned material that falls off the end of the grate; dust is the ash that falls from the combustion gas in the boiler passes; and fly ash is the ash that is removed in the pollution-control device. Siftings and bottom ash are usually captured in a water pit, which both cools the ash and seals the furnace. Dust and fly ash are removed dry from hoppers, although they are often mechanically conveyed to the water pit to be combined with the bottom ash and siftings for disposal.

Mass-burning plants are the most common type of thermochemical processing plants. More than 400 plants worldwide are currently recovering energy. Of these, 16 are operating in the United States. Seven RDF-dedicated boiler plants are operating (six in the United States and one in Canada).

Dedicated-RDF systems also have corrosion and erosion problems. In addition, if the temperature on the grate is not carefully controlled, aluminum and glass may melt and resolidify on moving parts; the correction of this problem does not seem to be well in hand, although limiting grate temperatures will help. Dedicated RDF units are also susceptible to slagging and development of deposits on the rear wall of the boiler; this problem can be corrected by controlling the air-swept feeder and by adding coatings and soot blowers on the rear wall.

A variation on the dedicated RDF boiler burns RDF normally as a partial replacement for coal in an existing spreader-stoker boiler. Test burns at up to 100% RDF by energy content have been successful; however, the boiler must be derated up to 15% of output in order to accommodate the higher moisture and ash contents of RDF. Problems include the somewhat higher particulate-matter rating on the pollution-control equipment (offset by lower sulfur emissions), the boiler derating and -- where the percentage of RDF is high -- overheating on the grate surfaces.

The suspension-fired boiler, designed to burn coal, is a third major variety of boiler for MSW-based fuels. Typically, utility boilers have been used because of their availability; a good grade of RDF fuel must be produced, and the ratio of RDF to coal is limited to 10% by energy content. The basic design includes an RDF plant located at the utility-boiler site. At the utility plant, some type of storage and reclamation facility (supplied by truck or conveyer) is provided. The RDF is reclaimed by a metering device, pneumatically transported, and blown into the boiler (usually through coal nozzles that have been repiped and converted for RDF). Generally, only half the RDF burns in suspension; the remainder falls to the bottom of the furnace. It is common practice for dump grates to be retrofitted in the furnace to hold the RDF until it completes combustion. The dump grates also tend to retain coal clinkers, which cool and are no longer brittle after they pass through the ash pit. This adversely affects the operation of the clinker grinders.

Another problem is strictly institutional. Most utilities obtain electricity from the cheapest source even if they must buy rather than internally produce the electricity. The plants converted to burn RDF are typically the older units. These units are operated for peaking power, so they are frequently off line. Meanwhile, MSW is made into RDF everyday and is not easily stored.

The last type of commercially operational MSW combustion unit is the modular combustor. There are more than 50 factory-erected small modular combustors, and 30 of them are in the United States. Modular combustors have many names and are made by a large number of manufacturers. They are characterized by assembly-line construction in a factory and they have one or more secondary chambers to allow for complete burnout of combustion gases. The typical unit (with energy recovery) has a capacity range of 30-120 tons per day. In principle the low velocity in the primary chamber and the fire in the secondary chamber should decrease and oxidize the particulate matter so that pollution-control equipment would be unnecessary. In practice, the units barely meet U.S. particulate-control standards, and many small plants include pollution control. Corrosion is not a problem in present-day modular combustors, because steam conditions have been limited to saturation conditions [temperatures of about 200°C (400°F)]. As the steam temperature rises, some corrosion occurs (as in the larger combustion units). Fouling can be a problem in these modular units, particularly if the boilers have obstructions. Other problems include refractory wear and combustion control.

A thermal system nearing commercial status is the fluidized-bed combustion system. Its bed of sand or other fine material is kept fluid by a rising current of air. Because of the large mass of the bed material and the inherent mixing of the air, RDF burns uniformly and rapidly. Bed temperature can be controlled precisely; if this temperature is limited to 815°C (1500°F), agglomeration of the bed is no problem because the glass does not soften. The amount of air necessary for combustion is controlled as much by the need to fluidize the bed as by the need to provide sufficient combustion air. Excess air can be limited to 45 to 50%, in comparison with 100% in convention burners. In addition, the bed material can be varied and used to capture chlorine and sulfur compounds. The first large-scale fluidized-bed combustor, built at Duluth, Minn., encountered fuel-preparation and fuel-feeding problems. The fuel preparation and fluid-bed configuration were modified and the system is in operation again.

Table 2 lists some operating thermal combustion facilities in the United States and Canada.

A number of commercial-scale MSW gasification systems have been constructed around the world. Four ANDCO-Torrax systems were built in Europe, one in Japan, and one in the United States. Of these six systems, two are operating, three are operational, and one has been dismantled. The ANDCO-Torrax system uses as-received MSW stored in a pit. A crane loads a lock hopper, which drops the MSW into a cylindrical chamber. As the waste falls, it is successively dried and gasified. At the bottom of the chamber the ungasified carbon char is burned with high-temperature air to provide a rising current of hot gas. The combustion-zone temperatures are hot enough to slag the metals, glass, and ash that flow from the unit through a slag tap. The hot gas produced, which has an energy content of 5.6 MJ/m^3 (150 Btu/ft^3) not including the sensible heat, is immediately burned in a secondary combustion chamber. Part of the hot gas is exhausted through regenerative towers, which heat the combustion air to 980°C (1800°F). The remainder of the gas then passes through a boiler and joins the gas from the regenerative tower before being exhausted through a pollution-control device.

Nippon Steel constructed a direct-heat system that uses preheated air enriched with oxygen. The facility, with a capacity of 450 tons/day, was constructed at Ibaragi City, Japan. It uses RDF, coke, and limestone. The end product, a fuel gas, has an energy content between that of the ANDCO-Torrax system and a Purox system.

The Purox system, developed by Union Carbide Corporation, is operating at Cichibu City, Japan. This system uses essentially pure oxygen to gasify RDF. The design is similar to the Andco Torrax unit. RDF is introduced through twin ram feeders. As the waste settles it is successively dried and gasified. At the bottom the char is combusted with pure oxygen. The slag from the bottom of the reactor is an obsidian-like glass that is stable and hard.

The Occidental Research system at El Cajon,

Table 2. Selected MSW Thermal Facilities

Facility type and Location	Year Operational	Design Capacity (TPD*)	Product
Field-Erected Mass Burning Plants			
Westchester County, N.Y.	1984	2250	Electricity
Chigago NW, Ill.	1972	1600	Steam for Industry
Saugus, Mass.	1975	1500	Steam for Industry
Montreal, Quebec	1970	1200	District Heat
Hampton, Va.	1980	200	Steam for Government
Gallatin, Tenn.	1981	200	Steam for Industry
RDF-Dedicated Boilers			
Dade County, Fla.	1982	3000	Electricity & Matls.
Niagara Falls, N.Y.	1981	2000	Cogeneration
Akron, Ohio	1978	1000	District Heat
Albany, N.Y.	1982	750	District Heat
Hamilton, Ontario	1972	500	Electric (1982)
Modular Combustors			
Pittsfield, Mass.	1981	360	Steam for Industry
Auburn, Me.	1981	200	Steam for Industry
Pascagoula, Miss.	1984	150	Steam for Industry
Susanville, Calif.	1985	96	Cogeneration
Red Wing, Minn.	1982	72	District Heat
Osceola, Ark.	1980	50	Steam for Industry
Groveton, N.H.	1972	24	Steam for Industry
RDF for Existing Boilers			
Baltimore County, Md.	1978	1200	Cyclone Utility Boiler
Ames, Iowa	1975	400	Suspension Utility Boiler
Madison, Wis.	1979	400	Industrial Boiler

*Tons per day.

Calif., was designed to use RDF in an entrained reactor to produce a liquid. The char was separated in a cyclone and burned to provide hot gas for the process. The liquid product was condensed from the gas stream exiting the reactor. The remaining gas stream (after the liquid product was condensed) was also used for system energy. The product liquid was highly oxygenated and viscous, and it deteriorated in storage. This system was shut down before some research questions were answered.

In Japan, two companies are developing indirectly heated fluidized-bed gasifiers. RDF is fed to a bed of hot sand, which is fluidized with product gas; the RDF gasifies. The bed velocity is high enough to elutriate the char, which is removed in a cyclone and burned in a second fluidized-bed to provide hot sand for the first bed. A similar system, using dual-media beds, is being developed in the United States for use with wood.

Pyrolytic systems produce oils, gases, and chars from RDF via thermochemical conversion in the absence or near absence of oxygen. These systems are in the research stage to increase our understanding of thermokinetic mechanisms, and to investigate improved methods of upgrading pyrolysis oils to liquid fuels. Slow and fast pyrolysis methods produce varying amounts of products, and with appropriate catalysts, high-quality liquid fuels should be achievable.

BIOCHEMICAL SYSTEMS

Although anaerobic digestion systems are probably the source of the natural gas we are recovering

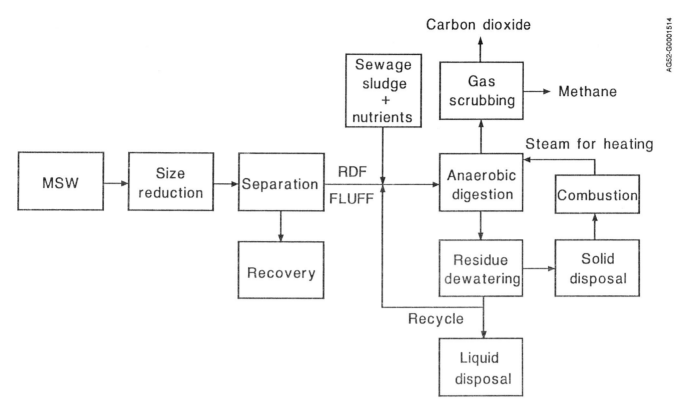

Figure 3. Anaerobic Digestion of MSW

today, and some have suggested that the anaerobe was the first form of life on earth, we have tried to control the process only since the late 1890s. The anaerobic digester (Imhoff Tank) was developed as a means to stabilize and treat sewage sludge. The biogas (a mixture of methane and carbon dioxide) was viewed as a waste and initially flared. By the 1920s, the gas was being used to produce power to operate sewage treatment plants. The productive use of biogas did not became a widespread practice in the United States. The environmental movement of the 1960s preferred combustion as the ultimate disposal means for sludge, and new construction for anaerobic digesters ceased. However, high energy costs and other concerns has awakened a new interest.

Biochemical systems use microorganisms or their enzymes to change MSW into fuel. Figure 3 charts the processes involved in anaerobic digestion of MSW to fuel. In controlled reactors, the MSW must be processed into RDF. An example is a reactor converting RDF to sugars for fermentation to ethanol. An uncontrolled reactor can process raw MSW; e.g., a sanitary landfill naturally degrading organics to methane and carbon dioxide.

A number of characteristics distinguish biochemical systems. One is the availability of oxygen to the system. Organisms are either aerobic (living only in the presence of oxygen), anaerobic (living only in the absence of oxygen), or facultative (living with or without oxygen). Another major distinction is whether the organism performs the conversion function as a part of its life cycle or produces an enzyme that catalyzes the conversion reaction. Compared to thermochemical systems, biochemical systems are slow (reaction times of hours or days), but they operate at mild conditions [near atmospheric pressure and below 100°C (212°F)] and produce specific products.

The major biochemical systems at an advanced research state are anaerobic digestion to produce methane rich gas and fermentation of cellulose to glucose for further conversion to ethanol.

Anaerobic digestion is a complex process that is not completely understood. Under controlled conditions, a series of organisms successively break down complex organic molecules into simpler molecules, ending with a mixture of half methane and half carbon dioxide. The process is thought to proceed in three stages. First, organisms or enzymes reduce the initial solids, which are complex polymers such as cellulose, fats, and proteins, to simple monomers such as sugar, fatty acids, amino acid, and other intermediates. This step is called fermentation and converts solids into soluble organic compounds. The biochemistry of this stage is not well understood. In the second stage, acetogenic bacteria reduce the products of the first stage to acetic acid, hydrogen, and carbon dioxide. In the third stage, methanogenic bacteria cleave the acetic acid to methane and carbon dioxide and reduce the hydrogen and carbon dioxide to methane and water.

The roles of the acetogenic and methanogenic bacteria are not well understood. Some of the organisms have been isolated but, in general, their biochemistry remains to be defined, and genetic manipulation is a project for the future. The acetogens and the methanogens have a very close interdependent relationship. Acetic acid and hydrogen are essential to the methanogen for its life processes, and, if these are not removed from the system, their accumulation will inhibit and soon stop the reactions of the acetogens.

The consortium of microorganisms is very sensitive to environmental conditions. While only the methanogens are strict anaerobes, they are vital to the success of the system and must be contained in an airtight reaction vessel. In addition, other factors such as neutral pH, proper nutrients (nitrogen, phosphorus, trace metals), absence of toxins, and appropriate temperatures are essential.

An important consideration in any biochemical system is the rate at which the active organisms double their population. For feedstocks such as sewage solids and manures, the organisms double and hydrolyze the substrate very rapidly. For the more recalcitrant substrates such as paper and wood, some believe this to be the rate limiting step. Once the substrate is hydrolyzed, the acetogenic bacteria grow rapidly and double their population in hours. The methanogenic bacteria grow more slowly with doubling rates of at least two days. If the substrate is applied (and therefore removed, since tank volume is fixed) faster than the doubling rate, the bacteria will be washed out of the tank and conversion will terminate. Thus a hydraulic residence time (HRT) of several hours permits the acetogenic organisms to reproduce but washes out the methanogenic organisms and the system would fail. In this case the methanogens control the process.

An anaerobic system has a series of controls. The process will proceed at any temperature from 5° to 60°C (40° to 140°F) with decreasing residence time to achieve the same level of digestion. The production of biogas is maximized at 37°C (98°F) and 60°C (140°F). Another important factor is retention time of the system. HRT is the time the average particle of liquid remains in the system, and solids retention time (SRT) is the time the average solid particle remains in the system.

A large fraction of MSW is biodegradable (See Table 1). However, certain organic molecules resist breakdown. These include the lignins that form 25% of wood and the majority of plastics and fibers. An appropriate feedstock preparation system must remove as much of this material as possible along with the nonbiodegradable inorganics. The measure of organic material in a substance is called volatile solids (VS). VS is determined by heating a sample to a controlled temperature for a controlled period of time. The loss in weight of the sample is the VS. Normally the sample is then heated until the weight is stable. The remainder is ash. Thus the nonbiodegradable portions of the sample contribute to the volatile solids. Other tests subject a sample to a controlled set of biological conditions. The loss of weight is the biodegradable solids.

A wide variety of types of reactor vessels can be used for anaerobic digestion processes. The simplest is similar to the domestic septic tank, which holds wastewater and in some instances garbage disposal solids. This type of reactor is uncontrolled and has a relatively short HRT and long SRT. Thousands of these simple units are in use, primarily in China and India, with manure as a feedstock; the biogas produced is used for light, heat, and as fuel for simple engines. The next most common is the single tank continuously stirred reactor (CSTR). This type unit is controlled and has an HRT equal to the SRT. It is common in a wastewater treatment plant and usually operates in the mesophilic temperature range. The biogas is used for heat and occasionally to operate engines for mechanical power and electric generation. Less common are systems with two or more tanks arranged in a variety of ways. The simplest scheme uses a CSTR followed by a tank that is not heated or stirred. The first tank quickly stabilizes the sludge, and the second tank retains sludge for an extended period for further digestion. The more complicated schemes have a small, first-stage CSTR where the substrate is converted to acid and a larger CSTR to complete the digestion process and produce biogas. The intent is to optimize the conditions for each of these two steps.

The most modern reactor being developed uses some form of a solid bed on which the organisms can be immobilized. Bed designs vary. A facility operating on distillery waste in Puerto Rico has plastic egg crate packing and the substrate flows upward through it. A 190-m^3 (6700-ft^3) sludge digester (ANFLOW) tested by the U.S. Department of Energy had plastic rings with the substrate flowing upward. Other schemes in development use horizontal or zigzag flow with plastic baffles and expanded or fluidized beds of sand. These units all have the same characteristics: very short HRT (8 hours or less) and very long SRT (up to a year or more). The short HRT adversely affects the energy recovery since as much as half of the methane can remain dissolved in the effluent, but the long SRT decreases the sludge disposal problems common with wastewater treatment plants.

No commercial MSW anaerobic digestion systems are currently operating, although there are numerous digesters operating on sewage sludge, primarily for stabilization. A large number of digesters using manure are operating, particularly in China and India. Research programs on MSW digestion continue in the United States, Italy, and Spain. A proof-of-concept refuse conversion to methane facility in Florida was built under contract to the Department of Energy. It was designed to accept 100 tons per day of MSW. Performance results did verify laboratory data, and a significant amount of information was collected from 1981 to 1985. Analysis indicated favorable economics for near-term commercialization of MSW anaerobic digestion systems. An experimental test unit at the Walt Disney World Resort Complex was designed to demonstrate biogasification concepts for methane production from biomass and wastes. Tests with RDF began in 1987; early results show 10%-50% increase in methane yield over CSTR digesters.

A second biochemical process converts MSW to ethanol. The ethanol can be used directly as an automotive fuel or as an automotive fuel additive, and there are many chemical uses in the industrial sector. Using ethanol and other alcohols as fuels is not new. Some countries such as Brazil require that all vehicles be fueled with pure ethanol.

Until recently, all industrial-grade ethanol was produced by thermal processes from natural gas. With oil price increases and grain surpluses in the United States, interest in fermenting ethanol from sugar and starch for fuel was rekindled. The increase in oil prices and the resultant serious balance of trade problem was sufficient to drive Brazil to produce ethanol from sugar cane and cassava starch.

With the rekindled interest in ethanol was a rekindled interest in also converting cellulosic-based feedstocks such as wood and MSW to ethanol. Conversion of both these cellulosic feedstocks is in experimental development, and proposals for pilot and commercial facilities have been made. While the majority of the research has been based on wood and agricultural wastes, this work is expected to be applicable to MSW because the major fraction of MSW is lignocellulosic material.

The MSW-to-ethanol conversion system has five steps: concentrating the components; pretreating the cellulosic and carbohydrate components; converting those components to sugars; fermenting the sugars to ethanol, and distilling the resultant ethanol/water mixture.

The concentration step has been described in the RDF preparation systems above. A potential complicating factor is removing plastics. A source separation system concentrating the food components might be particularly advantageous in a MSW-to-ethanol system since those tend to be higher in starches and sugars.

Pretreatment separates the lignin and cellulosic components, allowing the crystalline cellulose to be converted. The conversion of the cellulosic components may be accomplished by two techniques: acid and enzymatic hydrolysis. The former was described in patents in 1880 and used in World War I to hydrolyze cellulose to sugar mostly for animal fodder. It is relatively quick with reaction times in the range of seconds to minutes. Unfortunately the acids then hydrolyze the sugars to undesirable products from an energy standpoint with reaction times on the same order of magnitude. Enzymatic hydrolysis was discovered in World War II in the South Pacific. The rapid deterioration of cotton fabrics was traced to a fungus that selectively converted cellulose to glucose. Neither of these techniques are used commercially, although research and development is being conducted on both. CADCO, a DOE-funded joint project of United Biofuels, Foster-Wheeler, Raphael Katzen Associates, Riddick Engineering, and the University of Arkansas, is investigating the relationship between pretreatment and enzymatic hydrolysis in simultaneous saccharification and fermentation processes using pilot or commercial-scale equipment.

The conversion of glucose and other six-carbon sugars is straightforward. Yeast break down the sugar, and produce ethanol, carbon dioxide and heat. The yeasts used function in the mesophilic range. Particularly in small systems, the lower temperatures are favored to reduce ethanol evaporation and the need for expensive recovery equipment at this stage of the process. The efficiency of the process depends on yeast strain, nutrient availability, sugar concentration, temperature, and potential contaminants. All of these can be controlled by known processes. The hydrolysis of the hemicelluloses produces five-carbon sugars such as xylose. These are not readily converted to ethanol, although potential yeast strains have been discovered and are in research. The eventual conversion of hemicellulose to ethanol will increase the system yield by 50%.

The final step in the process is to separate the ethanol from the water and any unreacted sugars, yeast, etc. Since ethanol boils at 77.7°C (172°F), the separation is accomplished by steam distillation. At an ethanol concentration of 96%, the combined water and alcohol mixture has a constant boiling temperature (azeotrope) and further distillation is not possible. Since the use of ethanol as an automotive fuel or fuel extender requires a pure ethanol, the water fraction is displaced with another liquid such as benzene, and the thermal distillation to pure ethanol is completed. Research is under way to develop new purification procedures, including the use of molecular sieves, membranes, reverse osmosis, and absorption of the water on corn meal.

The net energy balance for a system converting starch to fuel-grade ethanol indicates that if oil is used to drive the process there is no net increase in liquid fuel supplies. The use of a renewable fuel or coal to operate the process is essential.

CONCLUSIONS

To produce energy from MSW and utilize this vast resource, we need to develop economic solid fuels as well as economic conversion processes to produce liquid and gaseous fuels. Technologies that exist today must be enhanced by improving conversion efficiencies and alleviating environmental concerns.

A better understanding of mechanical processing systems and the interdependence of basic unit operations is also required. Systems for making densified RDF and powdered RDF economically that can be easily transported and stored and are environmentally acceptable for use in conversion processes must be further developed.

Controlling environmental emissions is a necessity for the success of mass-burn systems. Additional research is required to increase understanding of the basic thermokinetic mechanisms so other thermochemical conversion systems can be further developed.

An increased understanding of anaerobic digestion systems will allow us to economically produce methane from MSW. Biochemical conversion processes need more research to improve their efficiency and the overall process economics.

Required research to make these options feasible is being sponsored through the Department of Energy's Biofuels and Municipal Waste Technology Division.

BIBLIOGRAPHY

Alter, H., *Materials Recovery from Municipal Waste*, Marcel Dekker, Inc, New York, NY, 1983.

Alter, H., and Dunn, *Solid Waste Conversion to Energy*, Marcel Dekker, Inc, New York, NY, 1980.

A Review of Comparative Energy Use in Materials Potentially Recoverable from Municipal Solid Waste, DOE Report No. DOE/CS/20167-12, 1982.

Assessment of Explosion Hazards in Refuse Shredders, ERDA-76-71, National Technical Information Service, Springfield, VA, 1975.

Diaz, L., et al., *Resource Recovery from Municipal Solid Wastes*, CRC Press Inc., Boca Raton, FL, 1982.

Energy from Biological Processes, Volume II-Technical and Environmental Analyses, Office of Technology Assessment, Suprintendent of Documents, US Government Printiing Ofice, Washington, DC, 20402, Stock No 052-003-00782-7.

European Waste-to-Energy Systems, HCP/M2103-0006, US Department of Commerce, National Technical Information Service, Springfield, VA, 1978

Geyer and Gammel, *Municipal Solid Waste Processing Systems Computer Model*, ANL-ENG-TM-03, Argonne National Laboratory, Argonne, IL, 1985.

Glasstone, *Energy Deskbook*, DOE/IR/05114-1, US Department of Energy Technical Information Center, Oak Ridge, TN, 1982.

Hainsworth, et al., *Energy from Municipal Solid Waste: Mechanical Equipment Status Report*, Idaho National Engineering Laboratory, Idaho Falls, ID, 1984.

Hasselriis, F., *Refuse-Derived Fuel Processing*, Butterworth Publishers, Boston, MA, 1984.

Hasselreis, F., *Thermal Systems for Conversion of Municipal Solid Waste, Volume 4, Burning Refuse-Derived Fuels in Boilers, A Technology Status Report*, Argonne National Laboratory, Argonne, IL, 1983.

Hickman, *Thermal Systems for Conversion of Municipal Solid Waste, Volume 1*, Argonne National Laboratory, Argonne, IL 1983.

Hopper, *Thermal Systems for Conversion of Municipal Solid Waste, Volume 3, Small-Scale Systems, A Technology Status Report*, Argonne National Laboratory, Argonne, IL, 1983.

Incineration, A State-of-the-Art Study, National Center for Resource Recovery, Lexington Books, Lexington, MA, 1974.

Institutional Issues Concerning Energy from Municipal Waste: A Status Report, Report No. ANL/CNSV-TM-123, Argonne National Laboratory, Argonne, IL, 1983.

Kuester, *Thermal Systems for Conversion of Municipal Solid Waste, Volume 5, Pyrolytic Conversion, A Technology Status Report*, Argonne National Laboratory, Argonne, IL, 1983.

Pfeffer, J. T., *R&D Data Base for System Development, RefCOM, A New Technique to Convert Municipal Wastes to Energy*, Argonne National Laboratory, 1985.

Pfeffer J. T., et al., *Biochemical Conversion of Municipal Solid Waste: A Technology Status Report*, ANL/CNSV-TM-122, Argonne National Laboratory, Argonne, IL, 1983.

Proceedings of the International Conference on European Waste-to-Energy Technology, Report No. ANL/CNSV-TM-14, Argonne, IL.

Reese, E. T., "History of the Cellulose Program at the U.S. Army Natick Development Center," *Biotechnology and Bioengineering Symposium*, No. 6, p. 9, 1976.

Rugg, "The New York University Process for Continuous Acid Hydrolysis," presented at the Solar Energy Research Institute Contractors Meeting, Golden, CO, 1982.

Steam, Its Generation and Use, Babcock & Wilcox Co, New York, NY, 1978.

The Recovery of Energy and Materials from Municipal Solid Waste, National Academy Press, Washington, DC, 1981.

Trezek, G. J., *Thermal Systems for Conversion of Municipal Solid Waste, Volume 6, Fluidized Bed Combustion, A Technology Status Report*, Argonne National Laboratory, Argonne, IL, 1983.

Turner, *Thermal Systems for Conversion of Municipal Solid Waste, Volume 2, Burning of Solid Waste in Large-Scale Combustors: A Technology Status Report*, Argonne National Laboratory, Argonne, IL, 1983.

Vesilind, P. A., and A. E. Rimer, *Unit Operations in Resource Recovery Engineering*, Prentice Hall, Inc, Englewood Cliffs, NJ, 1981.

Walter, D. K., "Anaerobic Digestion of Municipal Solid Waste to Produce Methane," Unpublished report of the US Department of Energy, 1985.

ORGANIC EMISSION STUDIES OF FULL-SCALE COFIRING OF PELLETIZED RDF/COAL

M. Poslusny, P. Moore and K. Daugherty* ■ University of North Texas, Department of Chemistry, Box 5068, Denton, TX 76201

O. Ohlsson ■ Argonne National Laboratory, 9700 South Cass Ave., Argonne, IL 60439

B. Venables ■ TRAC Laboratories, Inc., Denton, TX 76201

During the summer of 1987 over 600 tons of pelletized RDF was cofired with coal at Argonne National Laboratories. Samples of fly ash, bottom ash and air emission samples from the combustion zone, after the combustion zone prior to the pollution control equipment, and after the pollution control equipment were collected. Preliminary analytical results on tetrachlorodibenzo-p-dioxins/furans, polyaromatic hydrocarbons and polychlorinated biphenyls from the collected samples are discussed.

*Author to whom correspondence should be addressed.

Introduction

The overall objective of this research effort is to investigate the emissions resulting from the combustion of Refuse Derived Fuel (RDF) from Municipal Solid Waste (MSW) and to reduce those emissions through the use of appropriate binders. Our group has been studying ways to accomplish the following: (1) to decrease the chemical and biological degradation rate of RDF: (2) to pelletize RDF through the use of inexpensive binders; and (3) to easily transport/store the pellets under ambient conditions for up to 6 months. However, perhaps the most important criterion for the success of the RDF material in realizing the goal of incinerating MSW will be to demonstrate the environmental safety of emissions resulting from its combustion. The principal binder that our group has developed for RDF pelletization is calcium hydroxide.

During June/July of 1987, over 600 tons of dRDF were blended with high sulfur coal in a 6 week continuous experiment at Argonne National Laboratory (ANL). The blends were then combusted in ANL's boiler #5, which is a spread stoker fired boiler and the only boiler which supplied ANL's energy needs during this time period. Various concentrations of calcium hydroxide (from 0 to 8 percent) and various concentrations of dRDF in the coal (from 0 to 50 Btu percent) were investigated. These tests were conducted with industry, state and municipality participation, both in the critiquing of the test plan, and witnessing the actual test runs.

Fuel & Blending

High sulfur Kentucky coal and dRDF containing 0,4, and 8 percent calcium hydroxide were blended at ratios of 10,20, 30 and 50 percent dRDF by heat input. Baseline data was obtained by firing 100 percent coal. The dRDF came from two different sources, Future Fuels, Inc., Thief River Falls, MN., and Reuter, Inc. Minneapolis, MN. The final blend ratios by Btu content of coal to dRDF for each of the twelve runs can be seen in Table 1.

During test runs, grab samples were taken at each of the two weight scales at two hour intervals and composited to enable determination of the percentage, by weight, of coal and dRDF. These fuel samples are being analyzed for: proximate analysis, ultimate analysis, heating value, ash fusion temperature and bulk density.

Boiler Configuration

The boiler plant at ANL consists of five boilers that provide the steam requirement for the entire laboratory. The boiler (#5) used in the study was built by the Wickes Boiler Company (currently Combustion Engineering Company) and was installed in 1965. It has a rated capacity of 170,000 lb/hour at a guage pressure of 200 psig saturated, which is equivalent to about 212×10^6 Btu/hour. At a maximum rated capacity of 170,000 lb/hour at a the coal use requirement, based on coal at 11,600 Btu/lb is 9.0 tons/hour, and at the average capacity of 85,000 lb/hour the coal use requirement is 4.1 tons/hour. The maximum firing rate used for boiler #5, equivalent to producing 130,000 lb/hour steam (output), is about 144.4×10^6. Further design performance data are shown in Table 2.

Air Pollution Control

The air pollution control equipment associated with Boiler #5, consists of a mechanical multiclone collector followed by a spray dryer absorber and a fabric filter baghouse.

The multiple cyclone which was manufactured by Western Precipitation, Inc. contains 105 cyclones and has a particulate removal efficiency of 90% for gas flow rates of 50,000 SCFM and 25,000 SCFM, respectively.

After the flue gases exit the multiclone collector they are ducted into the spray dryer absorber (SDA) through a system of two gas dispersers. While passing through the two dispersers, the flue gas contacts a fine spray of absorbent (lime feed slurry). Sulfur dioxide (SO_2) is adsorbed into the alkaline droplets as water is simultaneously evaporated. In this way, the flue gas is cooled down. Control of the gas distribution, lime feed rate, temperature, and pressure within the SDA module assures that the reacting droplets reach their desired dryness before they leave the SDA chamber. A portion of the dry product, consisting of flyash, calcium sulfite, calcium sulfate, and unreacted lime, falls to the bottom of the absorption chamber. The material is then conveyed to the recycle material disposal silo. The SDA is designed to remove a minimum of 78.3% of the SO_2 contained in the flue gases exiting the boiler-multiclone system when the total boiler flue gas flow rate varies from 87,600 pounds per hour (35% MCR) to 219,000 pounds per hour (100% MCR). Design performance data for the SDA are given in Table 3.

The treated and cooled flue gases exhaust from the SDA module and flow to the baghouse. Here, the remaining fly ash and entrained spent dry chemicals are removed. The clean scrubbed flue gases exit the baghouse and pass through the inducted draft fan and stack, exhausting to the atmosphere. Performance design data are shown in Table 4.

On-Site-Sampling

Throughout the test our group took the following samples: coal, dRDF, bottom ash, fly ash, lime and ash, lime, and manual emissions. Manual emission samples included organics, metals and acid gases. Also, the emissions data used for pollution control in the daily operation of the plant was made available to us.

Our main consideration here will be with the organic samples; in particular polychlorinated biphenyls (PCB's), polyaromatic hydrocarbons (PAH's); dioxins and furans. These samples were collected at the sites listed below. The combustion zone was located at the center of the boiler which was considerably cooler than the back of the boiler where temperatures exceeded $2000°F$. The boiler configuration prevented sampling at the back of the boiler.

Site #		Approximate Sampling Temperature °F
1	combustion zone	1200
2	prior to the SDA	320
3	after the pollution control equipment and prior to the exit to atmosphere	170

An EPA modified method five sampling train was used for the collection of all organic samples. The probe and particulate filter were maintained at a temperature of $225° - 250°F$. XAD-2 resin was used to trap the majority of the organics. The first two impingers each contained 100 mL of water, the third impinger was empty, and the last two each contained 250g of silica gel. The final impinger temperature was kept below $70°F$.

Results and Discussions

Dioxin and furan analysis, of selected ash samples were conducted. We studied composite samples so that the results would be representative over longer periods of time in which the combustion tests took place. In addition, we decided to investigate only fly ash and not bottom ash, because fly ash would be expected to have higher concentrations of these organics.

Thus, we investigated 5 different fly ash periods as described in Table 5. These periods were during June 21 and 22; July 7; June 15 and 16; June 23 and 24; and July 4 and 5, results given in Table 6.

The major problems in the analysis of the samples has been the large concentrations of interfering species which are due to the "dirtyness" of the samples. PCB and PAH analysis has been possible because of the relatively large amounts of these species present in the sample. However, dioxin and furan levels are expected to be present at much lower concentrations, if at all.

The dioxin and furan data were taken from extracted emission samples run by gas chromatography-mass spectroscopy (GC/MS). The data include the tetra-chlorinated dioxins and the tetra-chlorinated furans per cubic meter of gas sampled. The final column lists the detection limit for each particular run and sample number.

The 2,3,7,8 tetra chlorinated isomers of dioxins and furans are by far the most toxic. Thus, we have initially focused on determining the levels of these isomers. The complete analysis for the tetra through octa congener groups will be undertaken after sample clean-up procedures have been optimized.

The results presented in Table 7 are those of the Site 3 samples analyzed for tetra chlorinated dioxins and furans. No tetra chlorinated dioxins or furans have been seen at this site. Since this site is after the pollution control equipment emissions are expected to be at a minimum here.

The PAH's that are being determined have been limited to the sixteen most hazardous, as listed by the EPA, see Table 8. All the congener groups of the PCB's are being studied. So far all fifteen Site 3 samples and fourteen Site 2 samples have been analyzed. The results are listed in Tables 9 and 10.

The binder has been found to considerably lengthen the storage lifetime of the pellets, its primary objective. However, it is too early to draw any conclusions on its effect on organic emissions. The only noticeable trends to this point are: Napthalene is the major component of PAH emissions and tetra and penta chlorinated biphenyls are the major components of PCB emissions.

TABLE 1

Coal/dRDF Run Breakdown

Run #	Composition	Binder %	Approximate Length of Run in Hours
1	Coal	-	89
2	Coal/10% dRDF	0	74
3	Coal/10% dRDF	4	74
4	Coal/10% dRDF	8	74
5	Coal/20% dRDF	0	112
6	Coal/30% dRDF	8	60
7	Coal/20% dRDF	4	68
8	Coal/20% dRDF	8	52
9	Coal/30% dRDF a	0	24
10	Coal/30% dRDF a	4	15
11	Coal/30%+ dRDF	4	30
12	Coal	-	114

a contained dRDF with reduced plastic

TABLE 2

Test Boiler (Boiler #5) Design Performance Data

Manufacturer:	Wickes Boiler Company (now Combustion Engineering Company)
Date Installed:	1965
Boiler Area:	17,647 FT^2
Economizer Area:	11,900 FT^2
Water Wall Area:	1,345 FT^2
Furnace Volume:	9,600 FT^3
Turndown Capability Ratio:	3.3 to 1
Firing Equipment:	Hoffman Stoker-Grates
Performance Based on Fuel of Not Less Favorable Analysis Than:	Bituminous Coal

Moisture	13.0%
Volume Matter	36.5%
Fixed Carbon	41.1%
Ash	9.4%
Total	100.0%
Btu/lb (as fired)	11,200
Fusion Temp. of Ash	2050°F

Load	1/2 MCR*	MCR*
Fuel	Coal	Coal
Steam Output, 10^6 lb/hr	85.0	170.0
Press, in Boiler Drum (psig)	200	200
Temp. Feed Water Entering Feedwater Heater	228°F	228°F
Temp. Feed Entering Econ.	292°F	270°F
Temp. Feed Leaving Econ.	350°F	347°F
Temp. Air Leaving Unit	100°F	100°F
Temp. Gas Leaving Boiler	500°F	582°F
Temp. Gas Leaving Econ.	331°F	346°F
Excess Air-Boiler Exit	37%	30%
Excess Air-Econ. Exit	38%	31%
Wet Gas at Boiler Exit, 10^6 #/hr	113.1	217.0
Wet Gas at Econ. Exit, 10^6 #/hr	114.0	219.0
Air Weight Entering Unit, 10^6 #/hr	100.0	191.0
Draft in Furnace, "H_2O	0.10	0.10
Draft Loss thru Boiler, "H_2O	0.27	1.00
Draft Loss thru Collector, "H_2O	0.65	2.20
Draft Loss thru Econ., "H_2O	0.33	1.20
Draft Loss thru Glass, Baffle, "H_2O	0.22	0.80
Draft Loss thru Flues, "H_2O	0.19	0.70
Draft Loss Total, "H_2O	1.76	6.00
Air Press. Loss thru Burners, "H_2O	-	-
Air Press. Loss thru Damper Ducts, "H_2O	0.20	0.70
Air Press. Loss thru Stoker, "H_2O	0.55	1.10
Air Press. Loss Total, "H_2O	0.75	1.80
Water Press. Loss thru F.W. Htr., #/in^2	2.1	8.2
Fuel Burned, 10^6 #/hr	8.99	18.00
Liberation, Btu/hr/FT^3	10,450	22,500

TABLE 2 (Continued)

	1/2 MCR*	MCR*
Heat Losses		
Dry Gas (%)	5.94	6.01
H_2 & Moist. in Fuel (%)	5.21	5.24
Moist. in Air (%)	0.15	0.15
Unburned Combustible **	1.30	2.30
Radiation	0.90	0.40
Manufacturer's Margin	1.50	1.50
Total Losses	15.00	15.60
Efficiency	85.00	84.40

*   Maximum Continuous Rating
** Based on 50% Recovery

TABLE 3

SDA Design Performance Data

Maximum Continuous Rating (% MCR)	35**	70**	100
Flue Gas to Absorber			
(ACFM)	28,869	53,677	73,608
(lb/hr)	87,600	153,300	219,000
Flue Gas Inlet Temperature to Absorber (°F)	330	346	346
Flue Gas Outlet Temperature from Absorber (°F)	146	148	148
Flue Gas ACFM to Baghouse	24,415	43,059	61,519
Flue Gas Inlet Temperature to Baghouse (°F)	146	148	148
Flue Gas Outlet Temperature from Baghouse (°F)	136	138	138
Flue Gas Dew Point Temperature Outlet from Baghouse (°F)	126	128	128
Flue Gas ACFM from Baghouse	24,620	43,415	62,035
Barometric Pressure (mm Hg)	760	760	760
Sulfur in Fuel *(%)	3.5	3.5	3.5
SO_2 from Boiler (lbs/hr)	441	882	1,260
SO_2 Leaving Bag Filter at 78.3% Removal (lbs/hr)	95.7	191	272
Outlet Loading Baghouse (GR/ACF)	0.01	0.01	0.01
Flue Gas Analysis (Volume %)	35% MCR	70% MCR	100% MCR
O_2	6.74	5.00	5.00
N_2	75.73	74.90	74.90
CO_2	10.32	11.82	11.82
H_2	6.98	8.02	8.02
SO_2	0.23	0.26	0.26
	35% MCR**	70% MCR**	100% MCR
Raw Pebble (lb/hr)	479	838	1,198
Water	11.6	19.4	26.4
Atomizer KW	50	82	111
Disposal Material (lb/hr)	1,074	1,970	2,814

* As specified

** These figures are calculated with the assumption that the flue gas flow in the 70% case is in direct proportion with the 100% figures supplied and that the flue gas flow in the 35% case is equal to 40% of the flow in the 100% case.

TABLE 4

Baghouse Design Date

	35% MCR**	70% MCR **	100% MCR
Gas volume (ACFM)	24,415	43,059	61,519
Temperature (°F)	146	148	148
Bag Fabric			Huyck Glass Fiber
Number of Compartments			4
Bags per Compartment			280
Bag Diameter (inches)			6
Bag Length (feet)			12
Total Filter Area per Compartment (sq. ft.)			5,278
Total Air-to-Cloth Ratio			
Gross	1.2:1	2.2:1	3.03:1
Net (with 1 comp. off-line)	1.6:1	2.8:1	4.04:1
Hopper Outlet Clearance to Ground			4 feet
Length			48.1 feet
Width			30.0 feet
Height			45.0 feet

TABLE 5

Analyzed Fly Ashes Compositions

Date	Composite Sample #	Composition
June 21-22	1	20% Btu Content Reuter RDF and 70% Btu Content Coal - 0% Binder
July 7	2	30% Btu Content Reuter RDF and 70% Btu Content Coal - 0% Binder (no plastic)
June 15-16	3	Coal Only
June 23-24	4	30% Btu Content Reuter RDF and 70% Btu Content Coal - 8% Binder
July 4-5	5	50% Btu Content Reuter RDF and 50% Btu Content Coal - 4% Binder

TABLE 6

Analysis of Composite Ash Samples for Dioxins and Furans

	Tetra Chlorinated Dioxins	Tetra Chlorinated Furans	Penta Chlorinated Dioxins/Furans	Hexa Chlorinated Dioxins/Furans	Hepta Chlorinated Dioxins/Furans	Octa Chlorinated Dioxins/Furans
Sample 1	BDL	BDL	BDL	BDL	BDL	BDL
Sample 2	BDL	BDL	BDL*	BDL*	BDL*	BDL
Sample 3	BDL	BDL	BDL	BDL	BDL	BDL
Sample 4	BDL	BDL	BDL	BDL	BDL	BDL
Sample 5	BDL	BDL	BDL	BDL	BDL	BDL
	LDL=1 ppb	LDL=5 ppb	LDL=2.5 ppb *LDL=5 ppb	LDL=2.5 ppb *LDL=5 ppb	LDL=2.5 ppb *LDL=5 ppb	LDL=1 ppb LDL=2.5 ppb

% Extraction Recovery Efficiency = 81%

BLD = Below Detection Limit

LDL = Lower Detection Limit

TABLE 7

Tetra-Chlorinated Dioxins

and Tetra-Chlorinated Furans at Site 3

Run#/Sample #	Site	Tetra-Chlorinated Dioxin Level	Tetra-Chlorinated Furan Level	Detection Limit
Run 1 Sample 1	3	BDL	BDL	0.72 ng/m^3
Run 2 Sample 1	3	BDL	BDL	1.99 ng/m^3
Run 2 Sample 2	3	BDL	BDL	4.07 ng/m^3
Run 2 Sample 3	3	BDL	BDL	5.24 ng/m^3
Run 3 Sample 1	3	BDL	BDL	4.80 ng/m^3
Run 4 Sample 1	3	BDL	BDL	4.27 ng/m^3
Run 4 Sample 2	3	BDL	BDL	4.27 ng/m^3
Run 5 Sample 1	3	BDL	BDL	0.49 ng/m^3
Run 5 Sample 2	3	BDL	BDL	0.47 ng/m^3
Run 7 Sample 1	3	BDL	BDL	4.16 ng/m^3
Run 7 Sample 2	3	BDL	BDL	4.10 ng/m^3
Run 8 Sample 1	3	BDL	BDL	4.78 ng/m^3
Run 8 Sample 2	3	BDL	BDL	4.78 ng/m^3
Run 12 Sample 1	3	BDL	BDL	3.85 ng/m^3
Run 12 Sample 2	3	BDL	BDL	4.85 ng/m^3

ng/m^3 = nanograms per cubic meter

BDL = Below Detection Limits

TABLE 8

EPA Priority PAH's

Napthalene

Acenapthylene

Acenapthene

Flourene

Phenanthrene

Anthracene

Fluoranthene

Pyrene

Benzo-a-anthracene

Chrysene

Benzo-b-fluoranthene

Benzo-k-fluoranthene

Benzo-a-pyrene

Dibenzo-a,h-anthracene

Benzo-g,h,i-perylene

Indendo-1,2,3-g,d-pyrene

TABLE 9

Polyaromatic Hydrocarbons (PAH's);
Polychlorinated Biphenyls (PCB's)
at Site 2

	Run #/Sample #	Site	mg PAH's / cubic meter of gas sampled	mg PCB's / cubic meter of gas sampled
	Run 1 Sample 1	2	1.7×10^{-2}	6.2×10^{-3}
a	Run 2 Sample 2	2	1.0×10^{-3}	1.3×10^{-2}
	Run 2 Sample 2	2	7.6×10^{-2}	2.7×10^{-1}
	Run 3 Sample 1	2	1.6×10^{-2}	1.4×10^{-2}
	Run 4 Sample 1	2	4.0×10^{-3}	7.6×10^{-3}
	Run 4 Sample 2	2	8.1×10^{-3}	7.7×10^{-3}
	Run 5 Sample 1	2	3.5×10^{-2}	9.7×10^{-3}
	Run 5 Sample 2	2	4.6×10^{-2}	7.7×10^{-3}
	Run 7 Sample 1	2	2.2×10^{-1}	2.0×10^{-3}
	Run 7 Sample 2	2	3.5×10^{-1}	2.9×10^{-1}
	Run 8 Sample 2	2	2.4×10^{-1}	1.3×10^{-2}
	Run 8 Sample 4	2	3.0×10^{-1}	3.4×10^{-3}
	Run 12 Sample 1	2	3.4×10^{-1}	5.4×10^{-3}
	Run 12 Sample 2	2	1.3×10^{-1}	$.3.9 \times 10^{-4}$

a This sample was lighter in color than all the rest.

TABLE 10

Polyaromatic Hydrocarbons (PAH's);
Polychlorinated Biphenyls (PCB's)
at Site 3

Run #/Sample #	Site	mg PAH's cubic meter of gas sampled	mg PCB's cubic meter of gas sampled
Run 1 Sample 1	3	4.6×10^{-3}	5.3×10^{-4}
Run 2 Sample 1	3	6.3×10^{-3}	1.2×10^{-3}
Run 2 Sample 2	3	1.5×10^{-2}	a
Run 2 Sample 3	3	8.1×10^{-3}	1.6×10^{-3}
Run 3 Sample 1	3	7.3×10^{-3}	9.1×10^{-3}
Run 4 Sample 1	3	7.3×10^{-3}	1.1×10^{-4}
Run 4 Sample 2	3	3.1×10^{-3}	3.1×10^{-3}
Run 5 Sample 1	3	3.6×10^{-4}	2.8×10^{-4}
Run 5 Sample 2	3	4.0×10^{-3}	1.2×10^{-3}
Run 7 Sample 1	2	7.9×10^{-2}	4.2×10^{-2}
Run 7 Sample 2	3	4.9×10^{-2}	6.5×10^{-3}
Run 8 Sample 1	3	1.0×10^{-3}	2.4×10^{-3}
Run 8 Sample 2	3	8.1×10^{-3}	8.5×10^{-4}
Run 12 Sample 1	3	7.0×10^{-2}	4.0×10^{-3}
Run 12 Sample 2	3	1.4×10^{-3}	4.3×10^{-4}

a Interference made it impossible to determine the quantity of PCB's in this run

ENVIRONMENTAL CONSEQUENCES OF MUNICIPAL SOLID WASTE DISPOSAL PRACTICES

Phillip J. Morris, Robert J. Bryan and Bruno I. Loran ■ Engineering-Science, Inc., 75 N. Fair Oaks, Pasadena, CA 91103

Local governmental officials face increasingly complex environmental issues and trade-offs when trying to decide what the appropriate method is for disposing of municipal solid waste (MSW). No method, within the bounds of technical and financial feasibility, is without some degree of environmental risk; and the public at large is becoming less tolerant of accepting these risks. The optimum systems of yesteryear or today, mass burn or landfill, are coming under glaring scrutiny as more is learned of their environmental consequences. Newer technologies and methods offer some clear environmental advantages, but are far from being environmentally benign.

The complexities of the MSW problem are becoming painfully apparent all over the United States. The Los Angeles basin is no exception. The recent collapses of LANCER and other incineration/power projects, as well as failed attempts at siting new landfills are reflective of a growing awareness of the environmental trade offs in MSW disposal options.

This paper examines the environmental consequences of MSW disposal options. It does not address disposal of industrial or toxic wastes. It concentrates on disposal, rather than pre-disposal collection and processing; and on air and water discharges and impacts. It is intended to provide a frame work for the papers presented and the panel discussions to be held this morning.

MSW DISPOSAL ALTERNATIVES

There are four basic steps in the MSW disposal process. These are illustrated in Figure 1. Collection at the source, households and commercial establishments, may reflect some degree of separation by the originator. For example, it is not uncommon for yard wastes to be separately collected from other wastes. Often the success of recycling or other disposal options depends upon source separation.

Intermediate processing may occur at transfer stations, landfills, resource recovery facilities, etc. Various intermediate processes may occur in various combinations. The processes used often depend upon predecessor intermediate processes. For example, shredding may occur before or after separation, and will impact the methods used for recycling.

The local vs remote options shown for landfills and incineration in Figure 1 reflect the pervasive influence of transportation costs on MSW disposal options. Serious attention is now being given to rail haul, in the western United States, of MSW or refuse

Table 1. Estimated toxic air emissions from selected landfill sites.

	Cedar Hills	Midway	Kent Highlands	Hidden Valley	Olympic View	Olalla	Hansville
Estimated steady state annual gas volume ($10^6 ft^3/yr$)	1679	480	1800	802	800	200	31
Estimated % through flares	37	82	88	10	90	0	0
Net Control efficiency (%)	33	74	79	9	81	0	0
Emissions of Toxic Components (Ton/Year)							
Toluene	4.85	0.15	2.27	2.80	0.58	0.77	0.11
Tetrachloroethylene	0.87	0.02	0.38	0.47	0.10	0.13	0.02
Methylene Chloride	0.16	0.02	0.97	0.69	0.14	0.19	0.03
Chloroform	ND	ND	ND	ND	ND	ND	ND
Trichloroethene	0.51	0.12	0.18	0.46	0.10	0.12	0.02
1,1,1-Trichloroethene	ND	0.00	0.00	0.00	0.00	0.00	0.00
1,2 Dichloroethlene	0.61	0.00	0.00	0.13	0.03	0.04	0.00
Benzene/CCL_4	0.29	0.50	2.75	2.81	0.58	0.77	0.11
1,1 Dichloroethene	0.09	0.00	0.00	0.02	0.00	0.01	0.00
1,1,2,2 Tetrachloroethane	ND	ND	0.96	0.62	0.13	0.17	0.03
1,2 Dichloroethene	0.00	0.00	0.00	0.00	0.00	0.00	0.00
Xylenes	0.93	0.30	1.26	1.59	0.33	0.44	0.07
	8.31	1.11	8.77	9.59	1.99	2.64	0.39

Source: Reference 2

derived fuel (RDF) to remote land fills or incineration facilities. This is in part due to the economics of rail vs truck hauling, the advantage of having an inplace rail system reaching almost every key locale in an urban area, and the attractiveness of the out-of-sight, out-of-mind syndrome.

Figure 1. Disposal options.

```
Source
    Mixed
    Separation
Collection
Intermediate Process
    None
    Separation
    Compaction
    Recycle
    Shredding
Disposal
    Landfill
        Local
        Remote
    Incineration With or Without Power
        Generation
        Local
        Remote
```

A comparison of two recent proposals for handling a portion of the 45,000 tons per day of MSW generated in the Los Angeles basin is shown in Figure 2. The project proposals have been normalized to 1,000 tons per day to provide a comparison which focuses on the volume of material which must be landfilled. This serves as a surrogate measure of environmental effects from landfilling and/or the siting of new landfills.

Figure 2. Comparison of proposed L. A. MSW disposal alternatives.

Proposed Los Angeles Recycle-Resource Recovery Project

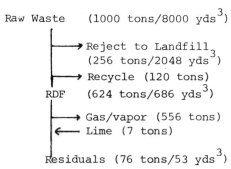

Proposed Los Angeles Recycle - Compaction - Landfill Project

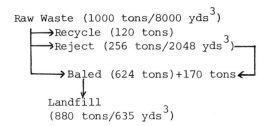

Volume landfilled=8% of volume delivered (635 yds^3)

The first proposed project, while similar to the LANCER proposals, incorporates a significant recycling process which generates a revenue stream potentially equal to one-half the revenue from sale of power. The second proposal is much less sophisticated, and the revenue stream, other than tipping fees, is totally dependent on recycling. The recycle-incineration project requires that twenty-six percent of the volume of the waste stream must be landfilled, whereas the recycle-compaction proposal requires only eight percent be landfilled.

This example illustrates the types of comparisons that can be made between disposal options. But let us delve more deeply into the environmental issues affecting MSW disposal, and some specifics of air quality and water quality impacts of the MSW disposal options of landfilling and incineration.

ENVIRONMENTAL ISSUES

The principal environmental issues confronted by proposed MSW disposal projects are illustrated in Figure 3. Water quality and water supply issues are linked by public health concerns over drinking water supplies. Otherwise, water quality issues deal with ecological impacts and other beneficial uses such as recreation. While landfills principally impact subsurface water quality, incineration/power generations mostly impact surface water supplies. Air quality impacts are currently the Achilles heel of incineration projects particularly in non-attainment air basins. Resource recovery, particularly recycling, is an environmental issue in its own right because of indirect impacts from recycling vis a vis product manufacture from virgin resources. Energy is of a similar nature because alternative sources of power most often require the consumption of non-renewable resources such as gas and oil.

Not-in-my-back-yard (NIMBY) is familiar to us all. The component environmental issues; traffic, noise, odor, etc. are often surrogates for the umbrella issue of compatible land use.

Figure 3. Environmental issues.

    Water Quality
    Air Quality
        Priority Pollutants
        Toxics
    Water Supply
    Resource Recovery
    Energy
    Siting/NIMBY
        Traffic
        Noise
        Safety
        Odor
        Dust
        Toxics
        Land Use

ENVIRONMENTAL EFFECTS OF LANDFILLS

Monitoring of gaseous emissions from MSW landfills has revealed significant quantities of toxics (See reference 1). Concentrations as high as 44 ppm of vinyl chloride, 32 ppm benzene, and 150 ppm toluene have been found. The origin of these compounds is not well known, although it is strongly suspected that vinyl chloride is a product of microbiological processes. Although the methods for estimating emissions from landfills is rudimentary, attempts have been made (See reference 2). Data given in Table 1 show a variation of a factor of about 25 between high and low values for total toxics emissions. This variation can be reduced to a factor of five if emissions are normalized to the same annual gas volume. Further reductions in the variation in quantities of emissions between landfills will be difficult to obtain, but will probably depend on age of landfill, climatic conditions, moisture content, etc. However, some estimate can be made as shown in Table 2. Gas production was assumed to be one cubic foot per pound per year (See reference 3); and emissions of toxic components totaled a rate of five tons per year based on the median of normalized values from reference 2. Reference 4 indicates somewhat different emission rates may be appropriate. This reference indicates that gas generation rates are on the order of 0.1 - 0.26 ft^3/lb/yr. The variation is attributed to "wet" vs "dry" climates. However, emissions of VOC's, not necessarily equivalent to the "toxic" components given in reference 2, which are given in reference 4, are apparently not influenced by climate; and are a fairly constant 7.5 tons/yr/$10^9 ft^3$ of gas generated. Therefore, the values in Table 2 may be high.

Table 2. Toxic air emissions from MSW landfills.

Capability	= 1,000 tons/day
Estimated Gas Production	= 730(10^6)ft^3/yr
Volatile Toxic Emissions	= 3.6 tons/yr
Components	
Vinyl Chloride	= 0.11 tons/yr
Benzene	= 0.76 tons/yr
Toluene	= 2.1 tons/yr
1,2-DCE	= 0.15 tons/yr
Miscellaneous	= 0.48 tons/yr

Reference 1 indicates that vinyl chloride is strongly suspected to be a result of biological action within landfills. If this is the case, vinyl chloride concentrations would be expected to be higher with the higher moisture contents which promote higher microbiological activity.

Of further interest is the observation based upon data and calculations given in reference 5, that landfills may be a major source of vinyl chloride in urban air sheds. The point source emission rate for vinyl chloride in the South Coast Air Quality Management District (SCAQMD) is 1.37 tons per year. Dispersion calculations for vinyl chloride, for an emissions inventory not including landfills, resulted in an annual average ambient concentration five orders of magnitude lower than measured. Landfills in the SCAQMD may account for as much as three to five tons per year. Clearly, landfills are candidate sources for major emissions of toxics in large urban areas.

Landfills also produce leachate. Rainfall is generally acknowledged as the major factor in leachate formation; however, MSW itself contains significant quantities of moisture. Reference 3 indicates that a good

Table 3. Toxic organics in groundwater from leachate production (units in part per billion, ppb).

Chemicals July	Well ID				State Action Level[a]	Gas Monitors[b]	
	MW-1	MW-2	MW-3	B-1		June	July
Vinyl chloride	N/A	ND	28	ND	2	4.0	4.4
1,2-Dichloro-ethylene (DCE)	N/A	7	42	ND	0.1 to 0.4	ND	ND
Trichloro-ethylene	N/A	4	3	ND	5	ND	0.8
Tetrachloro-ethylene (PCE)	N/A	64	12	ND	4	NA	NA
1,2- & 1,4-Dichloro-benzene	N/A	ND	12	ND	130	ND	1.3
Dibutyl phthalate	NA	ND	3	ND	350	NA	NA
4-Chloro-3-Methylphenol	N/A	ND	5	ND	***	NA	NA

[a] State Action Level as defined in California Administrative Code, Title 22.

[b] Parts per million from separate gas monitoring wells.

N/A = Not available due to dry well.

NA = No analysis

ND = Non-detectable

NL = Not listed

Source: Reference 7

average is twenty percent although ten percent is quoted in reference 6 for a Southern California landfill. For our 1,000 ton per day example, this translates to about 6,400 ft^3 of fluid contained in one days waste. Assuming one day equals one cell in a landfill placed in four foot lifts at 800 lbs per cubic yard density, a filtration rate of 30 gallons per day can be estimated using a 0.022 gal/day/ft^2 rate for silty clay substrate. Leachate contains perhaps 0.1 percent dissolved organics which translates to a rate of about ten or twelve tons per year infiltrating out of a landfill accommodating 1,000 tons per day. These are, of course, rough estimates, and can be greatly influenced by site specific geotechnical considerations. In addition, chemical and biological actions take place as leachate travels through soil which may capture and/or transform dissolved metals and organics. Table 3 abstracted from reference 7, provides a listing of some of the components found in groundwater affected by leachate from a municipal landfill.

This landfill in San Luis Obispo County, California, has been accepting municipal wastes since 1959. It now contains about 300,000 tons (500,000 yds^3). Table 3 also shows results from samples of landfill gas. Groundwater samples analyzed for non-organic constituents did not show any unusual levels.

If a leachate collection system intercepts landfill leachate, simple biological treatment systems can greatly reduce the amount of toxic organics, and facilitate the disposal of toxic residuals in an acceptable, controlled manner.

EFFECTS OF INCINERATION

Emissions of concern from MSW mass burn incinerators are dominated by metals, products of incomplete combustion (PIC's), and those suspected products of combustion dioxins (PCDD's) and furans (PCDF's). A fairly comprehensive analysis of emissions from such a facility is found in reference 6. The proposed project was a 1,034 ton/day mass burn facility involving separation, recycling, and the Best Available Control Technology (BACT) of air pollutants; it was proposed for the City of San Marcos, San Diego County, California. The projected emissions are shown in Table 4.

It is apparent that, in a gross sense, total emission of toxics from the incineration process for the San Marcos facility exceed those shown in Table 2 for air emissions from a landfill. A more equitable comparison would probably result if health risk assessments were available for each alternative. Unfortunately, such a comparison would be strongly influenced by the results for dioxins and furans since ambient concentrations derived from such low emission rates can exceed standards, as was the case for the San Marcos facility for PCDF in a non-residential area near the proposed plant site.

Table 4. Air emissions from a nominal 1000 ton/day mass burn incinerator.

Compound	Emissions, t/yr
Mercury	0.58
Lead	2.0
Beryllium	$2.6(10)^{-4}$
Fluorides	5.6
Cadmium	0.17
PCDD	$4(10)^{-4}$
PCDF	$4(10)^{-4}$
PAH	0.04

Incineration plants, particularly those with power generation, generate significant quantities of waste water, sludges and ash. Although ash is frequently exempted as non-hazardous, wastewater and sludges may not be so. In the case of the San Marcos facility, problems were avoided by a proposed ocean discharge, through an existing wastewater treatment plant outfall, of about 1,500 tons/day of blowdown streams, boiler water treatment sludges, etc. The sludge waste stream from boiler water treatment, hazardous if heavy metals are present, would have amounted to 36 tons/day.

VARIATIONS ON A THEME

Recycling is a highly attractive concept. It offers savings in energy and use of virgin resources, and some small but potentially significant revenue stream. Theoretical percentages of recyclables, including recycled yard waste, frequently exceed fifty percent. However, the price paid by users of these recyclable materials, aluminum, ferrous metal, paper, corrugated paperboard, plastics (film and bulk), etc. is highly dependent on quality and cleanliness of the product. Therefore, careful sorting and/or cleaning must take place either at the source, at the recycling center, or at point-of-use. At highly mechanized, centralized recycling operations, large quantities of wastewater are generated and treated or must be discharged to a publicly owned treatment works. The plastics content of MSW has more than doubled in the last ten years so that recovery of plastics is economically attractive. Its value rivals the revenue generated from aluminum. Processing of recovered plastics into new products creates additional volatile emissions which are not accounted for in most analysis of alternatives MSW disposal methods.

Compaction of MSW before landfilling is not commonly practiced, but is receiving more attention as haul distances to remote landfills becomes a more important element of overall MSW disposal costs. Compaction also plays a role in rail haul schemes as costs are sensitive to weight-volume ratios. Compaction may be antagonistic to recycling because as reference 7 suggests, a significant portion of the structural strength and integrity of compressed bales is the result of crushed metal waste. Compaction, either before or after recycling, offers some interesting considerations. For purposes of this discussion, compaction to a density of 1600 lbs/yd^3 is assumed compared to the long term, in-place compacted density in conventional landfills of about 800 lbs/yd^3. Table 5 lists some of these considerations taken from or based on reference 8.

TRENDS AND UNCERTAINTIES

Our knowledge and ability to quantify the environmental consequences of MSW disposal options is far from adequate. Additional information that is emerging from current research and risk assessments is forcing many local officials and regulators to face unpleasant facts, some of which are illustrated in this paper. Household waste contains toxic and precursors of toxic compounds which can be released to the soil, air or to surface and ground water.

Regardless of the state of our ability to predict consequences of MSW disposal options, there are two major areas of uncertainties we must face. First, we cannot predict the composition of the waste stream with sufficient certainty. The composition of our MSW today is different than it was ten or twenty years ago. This becomes very important for an incinerator/power plant whose technology will have to cope with this uncertainty for the economic life of the plant. New materials, packaging concepts, and products are introduced by the hundreds every year, and exert an influence on the composition of MSW.

The second major uncertainty is determining what is a safe practice. Most of the toxic substances we are now concerned with were either not known, or their toxic effects not appreciated ten or fifteen years ago. What assurance can we give our constituents that new knowledge about newly identified compounds or toxic effects won't render unsafe what is considered a safe practice today? These uncertainties are prompting much of the public resistance to proposed MSW disposal facilities.

Table 5. Environmental considerations of compaction.

Compaction process produces "squeezings" equal to about five percent of refuse weight. This reduces available moisture for biochemical reactions.

Permeability of compacted waste is reduced by a factor of eight to ten, reducing or eliminating leachate production.

Structurally stable for long distance track or rail transport.

Formation of organic acids in three to four days within bales essentially eliminates gas production.

Landfill volume required for given tonnage of waste is reduced by a factor of about two.

CONCLUSIONS

There appears to be four main conclusions that can be drawn from current information and trends which this paper has touched on. They are:
1. Sanitary landfills in the future will be highly engineered with leachate and gas collection and treatment systems.

2. Current mass burn and RDF incineration technology will give way to highly flexible and sophisticated approaches using high temperature and long residence times to assure complete destruction and pollution control.

3. Source separation, principally of toxic substances, will become essential to reduce contamination of the MSW stream.

4. Recycling will continue to struggle for

a secure niche in MSW disposal schemes, principally because of weak economic pressures and markets.

LITERATURE CITED

1. Wood, J.A., Porter, M.L., "Hazardous Pollutants in Class II Landfills", Hazardous Waste Management, Vol. 37, No. 5, pp. 609-615, May, 1987.

2. Engineering-Science Determination of Air Toxic Emissions From Non-Traditional Sources in the Puget Sound Region", prepared by Engineering-Science, Inc., for the USEPA Region 10 and the Puget Sound Air Pollution Control Agency, April 1986.

3. Tchobanoglous, G; Theisen, H.; Eliassen, R. Solid Wastes: Engineering Principles and Management Issues, McGraw-Hill, 1977.

4. Vogt, W.G., Conrad, E.T.; "VOC Emissions from Solid Waste Landfills", paper presented at Waste Tech, October 1987, San Francisco.

5. Shikiya, D; Liu, C; et al; "The Magnitude of Ambient Air Toxics Impacts From Existing Sources in the South Coast Air Basin", 1987 Air Quality Management Plan Revision Working Paper No. 3, South Coast Air Quality Management District, June 1987.

6. City of San Marcos, "Final Environmental Impact for the North County Recycling and Energy Recovery Center", Volume V, October 1984.

7. Peters, T; Makdisi, R; "Comprehensive Engineering Report Los Oso Landfill", prepared by Engineering-Science, Inc. for Engineering Department, County of San Luis Obispo, California, July, 1987.

8. Anon, Solid Waste Management Technology Assessment, prepared by General Electric Company, published by Van Nostrand Reinhold Co., 1975.

ENVIRONMENTAL CHARACTERIZATION OF REFUSE DERIVED FUEL INCINERATOR TECHNOLOGY

Robert E. Sommerlad and W. Randall Seeker ■ Energy and Environmental Research Corporation, Irvine, CA
Abe Finkelstein ■ Environment Canada, Ottawa, Ontario, Canada
James D. Kilgroe ■ U.S. Environmental Protection Agency, Research Triangle Park, North Carolina

The increasing cost and complexity of landfilling municipal solid waste (MSW), combined with the difficulties of locating new sites and expanding existing ones, has forced municipalities in both Canada and the United States to seriously consider available alternatives. The significant volume and weight reduction, and the energy potential, have made incineration of MSW an attractive alternative.

Current predictions by the EPA indicate that a substantial growth in municipal solid waste incinerators will occur over the next 10-20 years. Today about 100 MSW incinerators burn about 4 percent of the annual volume of MSW generated in the United States, whereas it is conceivable that by the year 2000 one-third of the MSW will be incinerated in more than 300 MSW incinerators. There is a definite trend moving toward incineration of MSW in the United States, and away from exclusively landfilling the waste.[1]*

In Canada, the majority of municipalities are facing a crisis in the disposal of increasing volumes of MSW. At the present time only 8 percent of municipal solid waste is incinerated with less than 4 percent in energy-from-waste (EFW) facilities. Less than 2 percent of Canada's municipal solid waste is recycled. In the next decade, up to 50 percent of Canada's waste could be disposed of in EFW facilities, with the construction of over 40 new facilities across Canada. This would provide a potential saving of 8 to 120 million barrels of oil annually.[1]

A concern raised repeatedly in connection with both existing and proposed MSW incinerators is their impact on health and the environment. Historically, poorly designed, controlled, and/or operated incinerators have resulted in nuisances and have demonstrated that environmental concerns can be sufficient to close facilities. More recently, the release of potentially toxic metal and organic emissions from these incinerators has become an issue.

In Canada, little or no development of MSW facilities will take place unless the public concern over the health and environmental impact of emissions from these facilities is satisfactorily managed.

---

*Numbers designate references.

To deal with these concerns, Environment Canada established the National Incinerator Testing and Evaluation Program (NITEP) in 1984. This five-year program is mandated: to identify the energy-from-waste technology most likely to be utilized in Canada; to assess relationships among state-of-the-art designs, operations, energy benefits and emissions; to examine the effectiveness of emission controls; and to provide input to National Guidelines for Emissions.

Under NITEP, three incinerator technologies have been identified for study and assessment. Two of the assessments have been completed to date. The first study was on two-stage modular incinerator technology suitable for small communities. The second study, on mass burning technology suitable for large communities, was completed in Quebec City in July 1986. The third test will be on a representative prepared refuse burning technology commonly referred to as refuse derived fuel (RDF).

The NITEP III field test program will be a joint United States EPA/Environment Canada evaluation. Of significant interest to Environment Canada is the need to complete the series of tests begun in 1984 under NITEP as part of its regulatory developments. EPA's interests lie in sharing some of the developments achieved under NITEP, as well as obtaining firsthand information on RDF incinerators for its regulatory requirements. In an effort to reduce the financial burden to both government agencies, a cost-sharing approach has been developed under the auspices of a Memorandum of Understanding signed between both departments in 1985.

The aim of this paper is to report on the progress of NITEP III to evaluate the current technology in the area of Refuse Derived Fuel (RDF) incinerator technology including the selection of a site and preparation of the combustion test program and how the test program will be conducted.

DESCRIPTION OF REFUSE DERIVED FUEL (RDF) INCINERATORS

The main difference between incineration of municipal solid waste in mass burning and refuse derived fuel (RDF) facilities is that in the latter case, the refuse is processed prior to burning. Processing can vary from simple removal of bulky items and shredding, to extensive processing producing a fuel suitable for co-firing in coal-fired boilers. RDF-based technologies also differ in the extent and form of both the fuel preparation, as well as in the combustion system. RDF does not necessarily have to be combusted on-site in dedicated boilers, but can be sold for firing as a fuel off-site. Some of the advantages of processing refuse are that it becomes more homogeneous and has a greater heating value per pound as well as requiring a smaller burning grate. Processing of the refuse prior to burning also allows a greater portion to be recovered for recycling.

Dedicated boilers used to combust RDF have basically the same design as those for coal-fired boilers, and can include suspension, stoker, and fluidized bed designs. If the RDF is sold as a fuel, it may be co-fired with a fossil fuel (usually coal). Dedicated RDF-fired boilers can be designed to handle over 1000 TPD. A typical schematic of an RDF processing facility with on-site boiler firing and recycling is shown in Figure 1.

RDF Technologies

RDF projects employ boiler systems manufactured by Combustion Engineering (CE), Babcock and Wilcox (B&W), Foster Wheeler (FW), Riley, or Zurn. CE and B&W combined represent approximately 80 percent (ton per day basis) of the active RDF projects. For the projects using B&W and FW boilers, a majority employ combustion systems supplied by Detroit Stoker (DS). The CE facilities use firing systems designed by CE, and Riley and Zurn use their own systems. To a large extent the Riley and Zurn systems are similar to the DS system.

PROGRAM

The overall program is divided into two phases with sub-phases as follows:

Phase 1. Part 1 - Site Selection
          Part 2 - Combustion Program.

Phase 2. Stage 1 - Mobilization and
                    Set-up

Stage 2 - Characterization Tests
Stage 3 - Performance Tests
Stage 4 - Data Processing and Reporting.

The product of Phase 1 will be used in developing a comprehensive test program on RDF incinerator technology. In addition to selecting the most appropriate site (Part 1), the program included setting preliminary test conditions, sampling locations and other relevant information necessary for designing a test program (Part 2). Two levels of testing will be employed in the program as follows:

° Characterization Tests

- Basis of understanding of the incinerator operating range, debugging of all systems, facility logistics and field crew familiarization.

  - 4-hour duration
  - Continuous Emission Monitors (CEMs).

° Performance Tests

- Extensive sampling and analysis, process evaluation and data evaluation.

  - 8-hour duration
  - CEMs
  - Sampling and analysis of organic emissions.

The specific parts of the first phase include:

Part 1 - Site Selection. A comprehensive review of all the RDF facilities in North America was conducted, their characteristics assessed and contacts were made with the owner/operators to determine whether they would participate in an evaluation program. After a ranking procedure had been established, five candidate sites were selected for review by the selection committee. After one site had been selected, a report was prepared describing the candidate unit and explaining the rationale for its choice.

Part 2 - Combustion Program. A document was prepared which will serve as a basis for setting preliminary test conditions, sampling locations, and other information which is necessary for designing a test program. The purpose of this test program is to develop criteria on the design and operating practices of RDF incinerators which will minimize emissions of toxic compounds without any adverse impact upon other emissions or the characteristics of solid residues. The test plan includes the following:

° Preparation of a host agreement.

° Resolution of economic impact of testing on the operation of the candidate unit.

° Preparation and documentation of a detailed test program which will assess all input and output streams as well as characterize the state of the unit.

° Development of test matrix test sheets on a spreadsheet format defining specific operating conditions.

° Identification of major activities for the field program.

° Establishment of two test series - characterization and performance.

° Contacting the manufacturer/builder of the unit to review the proposed tests.

° Establishment of all sampling requirements.

° Preparation of comprehensive cost estimates.

° Development of a process description of the test unit.

SITE SELECTION

The activities leading to the site selection included the following:

Available RDF Facilities. The first task in the site selection part of the program was to identify all RDF incinerators in North America. A list was compiled of all RDF units which are currently in operation, undergoing modification, or will be operational by 1989. The list amounted to 35 plants. Almost all of the units that were operational in July 1987 utilized electrostatic precipitators. One plant had been retrofitted with a spray tower and fabric filter. Several plants scheduled for operation in the 3rd and 4th Quarters of 1987 are equipped with spray dryers and fabric filters.

Ranking Criteria. In order to select the best site to host the test program, a list of ranking criteria was developed. The characteristics of each of the candidate units were evaluated using standard guidelines to avoid any biasing which could occur without such a system. The criteria were separated into three separate categories: Modern Practice, Flexible Operation, and Additional Considerations. A list of the ranking criteria is specified in Table 1.

The unit selected must be representative of the most modern design, since the purpose of the tests is to characterize plants which exhibit current and projected future technological trends in design and operation. The specific areas to be ranked in the category of modern practice include RDF preparation, RDF feeding system, stoker and grate, boiler/combustion system, and air pollution control devices. Each of these areas were given equal weighting of 20 points for a total of 100 in the first category.

The second category helps to evaluate the facility in terms of the flexibility of operation. The most important criterion in this category is the willingness of the owner/operator for a weighting of 50 out of 100 points. The cooperation of both the owner and operator is essential in any field test. It is also important that the permit requirements allow testing of the unit in the proposed manner.

The other minor areas of interest are the ability of the plant to accept RDF from other units, the number of units the plant has, and the plant set-up in terms of available space for trailers and test equipment.

The final category was the additional consideration of various factors. It is quite important to have adequate access for sampling of RDF, of the boiler, and of the stages in the Air Pollution Control Device (APCD). This area was given a weighting of 40 out of the final 100 points. Other areas of interest in a test program would include the control of the operation, whether or not the plant fires RDF which is representative of a "typical" unit, and the availability of the engineering data. A final consideration must be the arrangement of the APCD equipment and whether it would be possible to arrange for pilot scale equipment to be tested.

Preliminary Review. The process of selection of a host site continued with an initial screening of the available RDF plants. Many units were eliminated due to fuel type, grate design or age of the unit. Other units which plan to be in operation in late 1988 or 1989 were also eliminated since they will not be completed in time for the test program schedule.

After a second review which was based quite heavily on new units and those employing the most recent technology, the possible candidate selection was reduced again. The site visitation was accomplished in three days, with the site selection committee visiting three candidates. Each member of the committee was provided with a notebook containing the trip itinerary, some information on the layout and processing of each plant, and a list of the ranking criteria for each plant.

Selected Site. The Mid Connecticut plant was selected; the plant was designed and built by Combustion Engineering.[2] An artist's version of the complete facility is shown in Figure 2. The new boilers are housed in an old boiler house from which the old boilers were removed. The facility is located along the Connecticut River on the South side of Hartford in the South Meadow area. The plant is currently in start-up and will be in commercial operation by June 1988. The anticipated fuel capacity is 2000 tons per day. The plant will have three boilers each producing 231,000 lb/hr (29.1 kg/s) of steam and 68.5 MW of power.

The RDF processing is located in a building which is separated from the boiler house, with the RDF being transported to the boiler on covered conveyors. RDF processing at this facility can operate at a full load of 100 tons/hr. on each of the two processing lines. Municipal solid waste (MSW) is received from trucks onto a tipping floor. The MSW is then inspected to remove bulky items and hazardous material before being transported to a flail mill. After bags are broken open, iron and steel are removed by drum type magnetic separators. The bulk of the waste is then conveyed to large rotary trommel screens, which allow non-combustible residue such as glass and sand to be removed. The second stage of the trommel separates out the combustible fraction which does not need further size reduction. Oversize material is conveyed to a hammermill shredder for final size reduction.

The CE steam generating units shown in Figure 3 are natural circulation welded-wall boilers with a 2½ inch (64 mm) OD tubine on 3-inch (76 mm) centers. The upper section of the furnace contains widely spaced screen panels which cool the gases below the ash fusion temperature before they enter the superheater. The superheater consists of a vertical two-stage pendant platen design which is located over the furnace's nose arch to protect it from direct radiation. Following the superheater is the economizer which consists of two horizontal banks of in-line tubes.

The fuel burning system consists of a CE Refuse Combustor Stoker and specially designed grate. The grate includes a self-cleaning key design to remove fused or clinkered ash during grate operations. Multiple undergrate air zones provide controlled air flow to ten areas of the grate. The fuel is conveyed from the RDF storage area to surge bins located at the front of each boiler. A vibrating pan feeder is used to feed the fuel uniformly to the stoker.

The overfire air system is separate for the coal and the RDF burning. The RDF system consists of four tangential overfire air windbox assemblies located in the furnace corners. Preheated air is admitted tangentially to form a vortex. Overfire air ports for the coal combustion are located on the front and rear walls. One row of ports is provided on the front wall and two rows on the rear wall. Thus, overfire air either in tangential or wall-mode, or a combination of both modes is possible.

The ash removal system consists of two streams which are later combined and stored for eventual storage. The first stream collects the bottom ash, economizer ash and stoker siftings. The second stream collects baghouse and air heater ash. After this has been conditioned in a pug mill it is combined with the first stream and transported to a three-sided storage bin.

The flue gas cleaning system consists of a lime-based dry scrubber followed by reverse-air fabric filter. The scrubber has a separate control for both the flue gas flow rate and the limestone slurry addition flow rate, which is governed by the SO_2 and HCl levels downstream of the scrubber and the gas flow rate from the boiler. The fabric filters have twelve compartments, each with 168 woven glass filter bags.

Some of the other major advantages to this CE system include the multiple plenum undergrate air and the variety of overfire air configurations possible. Some of the key parameters to be examined in the testing include the various arrangements of overfire air and the impact which segregated undergrate air has upon emissions levels. Another significant factor which has a major impact on the decision of which facility will host the test program is the willingness of the owner/operator to cooperate. Combustion Engineering and the other owner-operator groups -- the Connecticut Resources Recovery Authority and the Metropolitan District Commission have been quite open in their desire to accommodate this program at their plant.

COMBUSTION PROGRAM

The activities leading to the combustion program are derived from previous NITEP combustion programs and include the following:

Requirements

The focus of the field combustion program for RDF units is on the following:

° Defining conditions where potentially hazardous pollutants (e.g., chlorinated dibenzodioxin (CDD)/chlorinated dibenzofuran (CDF) are formed (both high temperature and post-combustion zones).

° Validate the principal hypothesis that "good combustion" yields low CDD and hazardous hydrocarbon emissions.

° Define criteria for "good combustion" including design, operation, verification and monitoring and generate information that will promote improvement in incinerator technology.

° Provide data that will create confidence in the public that RDF incineration does not impose a serious health hazard.

The program plan must address these issues and be structured in such a manner to allow these goals to be achieved. A serious of preliminary "combustion" design and operating guidelines have been defined for RDF units that can serve as the hypothesis for the test plan.[1] These combustion practices are shown in Table 2. The test plan will specifically address each of these elements as thoroughly as possible. In addition, the test plan will address the critical issues of the location of CDD/CDF formation and destruction. Without this information, it will be difficult to develop advanced control schemes to minimize the emissions of CDD and related organic species. Appropriate measurements in full scale equipment will supply vital information on mechanisms that cannot be obtained reliably in any other manner. Finally, the combustion conditions which minimize the emissions of CDD and related trace organics can have an adverse impact on the formation of other pollutants - notably NO_x and metal fume. The impacts of proposed CDD control on these measures will be quantified and options to minimize the impacts evaluated.

A key equipment for the combustion program is to use the available money in an efficient manner to maximize the progress towards the technical goals. The planning phase will consider the resources and weigh the relative value of different activities on the basis of cost and technical reward.

The information generated in this test program will be specific to a single unit - the unit on which the testing is taking place. The data generated will undoubtedly be incomplete in terms of variations in parameters and detailed characteristics due to restrictions placed on the test from permits, operators and resources available. It is crucial that the data are interpreted and extended to be able to assess the full range of parametric effects on the test unit. Also, the data must be generalized to other units to develop combustion design and operating conditions which apply to all RDF incinerator systems. Only in this way will the data be useful for development of design and operating criteria that are generally applicable. Thus, the test planning phase will define an approach that will not only obtain performance data but that will allow the date to be interpreted and generalized.

Technical Approach

In Figure 4 is shown the procedure that has been followed in developing the characterization and performance test plan. In this part of the program, the characterization plan will be finalized. The performance plan will be defined but not finalized until the characterization tests and engineering analysis on these tests have been completed. This will ensure that the performance tests are of maximum utility for the development of design and operating criteria since the plan will be based on preliminary analysis of the criteria when applied to this unit.

One aspect of the program is to assess how the Good Combustion Practices that were developed by EER for RDF units apply specifically to the test unit. The criteria were translated to this unit to define the recommended conditions for minimizing emissions of trace organics such as CDD/CDF. These preliminary criteria represent the hypothesis that will be evaluated in the test program. The engineering assessment also included an evaluation of the current understanding of CDD/CDF formation/destruction in RDF plants and an identification of the potential locations of formation in the test unit. The mechanisms included the major mechanisms proposed in the literature to date, including:

° Trace oganics in the feed which are not destroyed.

° Formation of CDD/CDF in fuel rich zones in the near grate region and failure to destroy these species in the furnace.

° Formation of precursors with similar structure to CDD/CDF in the fuel rich zones in the near grate regions and conversion in the furnace by a chlorine donor.

° Formation of CDD/CDF on fly ash by a catalyzed reaction from precursors.

The characteristic temperatures for the mechanisms were identified and compared to the temperature profile within the test unit in order to identify candidate the temperature profile within the test unit in order to identify candidate sampling positions for evaluating the validity of the different mechanisms.

The characterization and performance tests were then defined based upon the results of the engineering assessment. The test plan serves to define conditions within the unit that can be used to address specific hypotheses based upon the mechanisms of CDD/CDF formation and the combustion criteria for preventing emissions.

Interpretation and Generalization of Field Tests

The focus of this program is to address the joint Environment Canada and US EPA objectives for the characterization and control of emissions from MSW incineration as shown in Table 3. This program will produce a test plan for RDF incinerator systems which will supply much needed information on the appropriate design, operation and monitoring of RDF incinerators to prevent or minimize the emissions of trace organics such as CDD/CDF. That which is desired is a <u>verified</u> design and operating guidelines that can be <u>generally</u> <u>applied</u> to RDF units and will ensure optimum control of all emissions. The data generated in the testing of the single RDF unit chosen in this program must be sufficient to generate the appropriate design and operating guidelines for all RDF incinerators.

REFERENCES

1. W.R. Seeker, W.S. Lanier and M.P. Heap, "Municipal Waste Combustion Study: Combustion Control of MSW Combustors to Minimize Emission of Trace Organics", prepared for the U.S. Environmental Protection Agency, Office of Research and Development, Washington, D.C. 20460, Report No. EPA/530-SW-87-021C.

2. M.D. Mirolli, W.B. Ferfuson, D.L. Bump, "Mid-Connecticut Resource Recovery Project," presented at the 1986 Joint Power Generation Conference, Portland, Oregon, October 19-23, 1986.

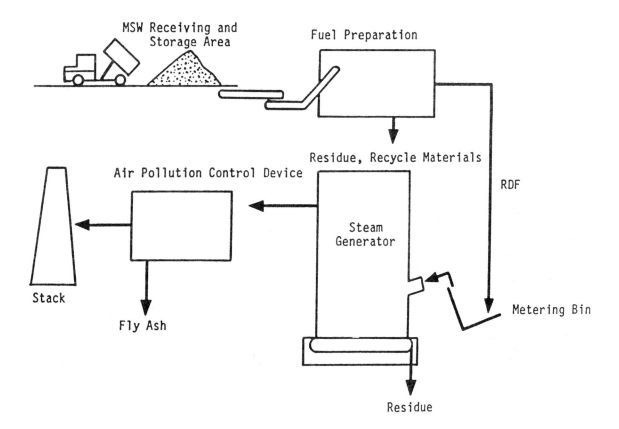

Figure 1. RDF processing facility with on-site boiler firing and recycling.

1. Refuse Truck Unloading Area
2. Refuse Shredders
3. Ferrous Metal Magnets
4. Primary Separation Units
5. Secondary Shredders
6. Metal Outloading
7. Residue Outloading
8. Secondary Separation
9. Refuse Derived Fuel (RDF) Storage
10. Refuse Derived Fuel (RDF) Conveyors
11. Barge Delivery of Coal
12. Coal Storage
13. Coal Reclaim Conveyors
14. RDF & Coal Conveyors to Boilers
15. C-E VU-40 RDF and/or Coal Fired Boilers
16. High Efficiency Emission Control Equipment (Dry Scrubber/Baghouse)
17. Stack
18. Turbine Generators
19. Switch Yard

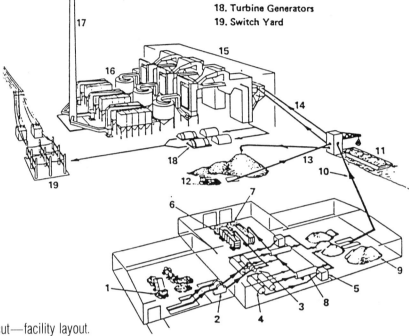

Figure 2. Mid-Connecticut—facility layout.

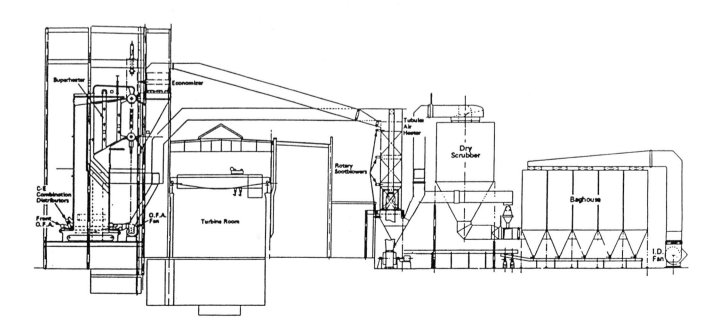

Figure 3. RDF steam generator and APC equipment for mid-Connecticut.

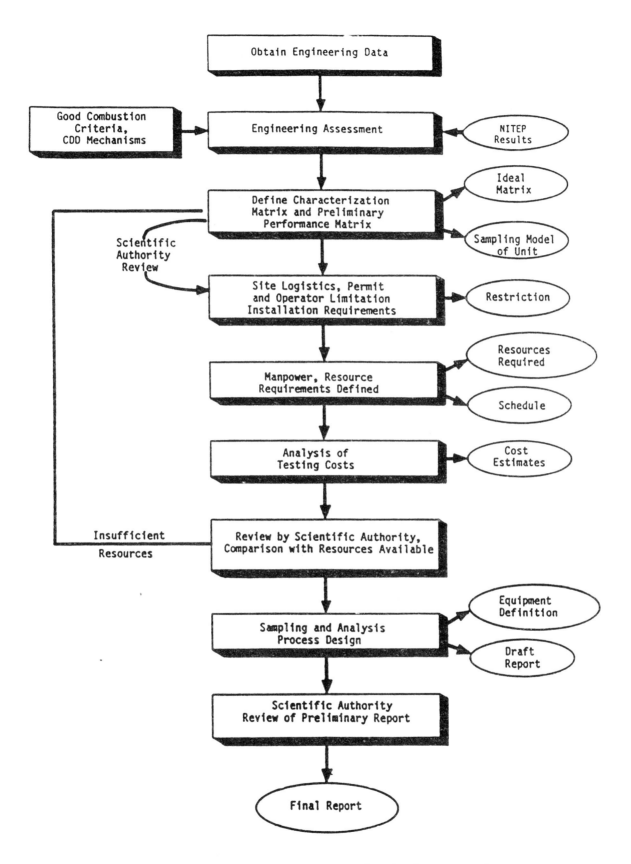

Figure 4. Technical approach to development of combustion program plan.

Table 1. Selection criteria.

Points		
	Modern Practice	
20	RDF Preparation	- Size Reduction/Classification - Metals Removal - Refuse Weighing Capability
20	Feeding Systems	- Controlled Uniform Feeding - Metered RDF Feeding
20	Stokers and Grates	- Design Representative of Modern Practice
20	Boiler/Combustion Systems	- Overfire/Undefire Air - Welded Walls - Refractory Cladding - Furnace Exit Temperature
20	APCD	- Ash Collection Capability - Type and Conditions
100		
	Flexible Operation	
50	Owner/Operator Willingness	
20	Permit Conditions	- Range of Operation - Ability to Modify
10	Accept RDF From Other Units	
10	Multiple Units	
10	Plant Set-Up	- Space - Access
100		
	Additional Consideration	
40	Sampling Access	- RDF - Number of Doors on Boiler - Flue Between Boiler & APCD - Stack Access - Plant Spacing
15	Control & Monitoring	- Metered Feeding - Btu Control System - Air Heaters - Plant Spacing
15	Representative RDF	
15	Availability of Engineering Data	
15	Spray Dryer/Baghouse Access for Pilot Scale	
-	Power Supplies	
-	Cost Sharing Possible	
-	Proprietary Information Identification	
100		
Total 300		

Table 2. Good combustion practices for minimizing trace organic emissions from RDF combustors.

Element	Component	Recommendations
Design	Temperature at fully mixed height	1800°F at fully mixed height
	Underfire air control	As required to provide uniform bed burning stoichiometry (see Ref. 1)
	Overfire air capacity (not necessary operation)	40% of total air
	Overfire air injector design	That required for penetration and coverage of furnace cross-section
	Auxiliary fuel capacity	That required to meet start-up temperature and 1800°F criteria under part-load operations
Operation/ Control	Excess air	3-9% oxygen in flue gas (dry basis)
	Turndown restrictions	80-110% of design - lower limit may be extended with verification tests
	Start-up procedures	On auxiliary fuel to design temperature
	Use of auxiliary fuel	On prolonged high CO or low furnace temperature
Verification	Oxygen in flue gas	3-9% dry basis
	CO in flue gas	50 ppm on 4 hour average - corrected to 12% CO_2
	Furnace temperature	Minimum of 1800°F (mean) at fully mixed height
	Adequate air distribution	Verification Tests (see Ref. 1)

Table 3. Joint program objectives.

- **National Incineration Testing and Evaluation Program**

    1. Define Optimal Design and Operating Characteristics

    2. Relate Operating Conditions to Emissions

    3. Identify Best Practical Control Options

    4. Design and Operating Guidelines for Future Applications

    5. Incorporate Accepted Dioxin/Furan Sampling and Analysis Protocol

    6. Investigate Dioxin Surrogate

    7. Facilitate Construction of New Incinerators

- **U.S. EPA Goals**

    1. Establish Baseline Emissions

    2. Compare Classes of MWC Systems

    3. Evaluate Design and Operating Parameters

    4. Evaluate Add-On Pollution Control Devices

    5. Establish Design and Operating Criteria

STATE-OF-THE-ART FLUE GAS CLEANING TECHNOLOGIES FOR MSW COMBUSTION

Theodore G. Brna ■ U.S. Environmental Protection Agency, Air and Energy Engineering Research Laboratory, Research Triangle Park, NC 27711

The control of air pollutants from the combustion of municipal solid waste (MSW) has evolved from none or only particulate control to current systems controlling acid gases, trace organics, trace heavy metals, and nitrogen oxides as well as particulate matter. Both in-furnace and post-combustion control methods for limiting one or more of these pollutants are discussed. Test results for various control technologies are reported. Recent control technology applications and trends are noted.

INTRODUCTION

Much design and operating experience on municipal solid waste (MSW) combustion has been gained in Japan and Western Europe over the past decade as the volume reduction of wastes has been stimulated by the declining availability and increasing cost of landfills. Nearly 2000 MSW units in Japan and several hundred in Western Europe are now operating, with the trend now being waste-to-energy conversion rather than simply incineration to reduce volume. Technology developed in Japan and Western Europe has been beneficial to the U.S., where over 100 MSW combustion systems are now operational and a similar number are in the construction or conceptual development phase.

The reduction of waste volume by combustion results in air pollution, including pollutants not currently regulated by the U.S. EPA. Table 1 shows the U.S. standards along with those of several states and countries (1). However, EPA has announced its intention to further regulate emissions from MSW combustors and proposes promulgation of these regulations in December 1990. Currently studies are underway to determine which pollutants to regulate and to what extent. As indicated in Table 1, classes of pollutants currently regulated by one or more of the entities listed include: trace organics (dioxins, total organics), acid gases (HCl, SO_2), trace heavy metals (Hg, Cd, Tl), and particulate matter. The listing in Table 1 is not intended to be complete. For example, West Germany regulates the emissions of more trace metals, and some U.S. states, as well as Japan and West Germany, have NO_x requirements/guidelines.

Noting the classes of pollutants that are currently regulated and their potential for regulation in the U.S.--on a national, state, or local level--the air pollution control strategy selected for a given plant should have the potential for multi-pollutant control, if costly retrofit or upgrading is to be minimized in meeting future regulations. Residues, although small in volume relative to unburned wastes, contain concentrated pollutants requiring environmentally safe disposition. Since flue gas cleaning is the primary objective here, residue disposal will not be discussed.

EMISSION CONTROL TECHNOLOGIES

In-Furnace Methods

The major concern of in-furnace emission control is the destruction of combustible and toxic components. As part of a recent EPA report to Congress, good combustion practice guidelines were stated (2). These guidelines, currently being reviewed in the regulatory development effort, address temperature control, overfire/underfire air mixes, turndown and start-up restrictions, CO and O_2 control, and other factors.

Limestone ($CaCO_3$) injection at a molar ratio of $CaCO_3$ to HCl of 1.2 into the furnace [700 to 950°C (1300 to 1750°F) zones] with HCl concentrations of 500 to 800 ppm is reported to reduce the HCl to 370 and 400 ppm and at a ratio of 4 to 100 and 130 ppm, respectively (see Figure 1). This method does not seem suited to a HCl removal requirement of 90% or outlet emission limits of 30 to 50 ppm, as required in some U.S. states, unless a more reactive sorbent is used. Figure 1 also shows the performance of in-duct injection of calcium hydroxide [$Ca(OH)_2$], but this will be discussed later under acid gas control in flue gases.

NO_x control methods as used in Japan are summarized in Table 2. Automatic combustion control, water spray into the furnace or combustor, flue gas recirculation, and ammonia injection listed in this table are in-furnace NO_x controls, while ammonia injection with a catalyst and a wet scrubber with oxidation are post-combustion NO_x control methods.

The first three methods control oxygen and temperature to limit NO_x formation. In the U.S., the thermal $DeNO_x$ method injects ammonia into the upper furnace to selectively reduce NO to N_2 (3). Since the ammonia/NO_x reactions are very temperature sensitive, the ammonia injection location must be carefully chosen (see Figure 2). Injection location can also affect ammonia slip (i.e., the ammonia that does not react with NO_x).

Reburning with an auxiliary fuel, such as natural gas or oil, is another NO_x control method and may be used in conjunction with flue gas recirculation. Its application to MSW combustors may be as shown in Figure 2. The stoichiometric ratio (SR_i) terms are for air/fuel mixtures, a value below 1.0 denoting a fuel-rich state. Sufficient fuel should be supplied at a low location in the furnace to produce a hot, oxygen-deficient zone. Overfire air injected over the reburning zone supports completion of the combustion process. Reburning can be combined with ammonia or urea injection to maximize NO_x reduction and also to promote destruction of organic compounds in the presence of high flame temperature and high concentration of flame radicals in the reburning zones (4).

Particulate Matter Control

The predominant method for particulate control in MSW combustion is the electrostatic precipitator (ESP). A low emission level [<45 mg/Nm^3 (<0.02 gr/dscf)] can be attained by an ESP with a high ratio of collector plate surface area to gas flow volume, such as 1700 m^2-min/Nm^3 (520 ft^2-min/ft^3) or larger (5).

Fabric filters are seldom used without upstream flue gas humidification (quenching) or sorbent injection to preclude bag failures caused by hot gases, spark carryover, or sticky particles from the furnace/boiler. They are capable of controlling particulate matter to below 45 mg/Nm^3 without operational upsets related to varying MSW and ash compositions that can adversely affect ESP performance.

Venturi scrubbers using water, used in early MSW incinerators, are not sufficiently effective to attain most current particulate control requirements as they remove 80 to 95% at normal operation. Very high pressure losses occur when venturis effectively remove fine particles. In addition, the high erosion and corrosion potential in venturis handling acid streams leads to costly operation and degraded reliability.

Acid Gas Control

Both dry and wet processes are used to remove acid gases (HCl, HF, and SO_2) from flue gas produced by MSW combustion. Here, dry and wet scrubbing means that the wastes from the process are dry and wet (slurry or sludge) products, respectively. In some countries, semi-dry or wet/dry scrubbing refers to a spray dryer (here an absorber because it receives an alkaline slurry or solution which is atomized into fine droplets that react with acid gases and are dried on contacting the surrounding hotter flue gas stream to produce dry powdery solids). Thus, the cooled flue gas leaving a spray dryer is unsaturated with water and carries entrained dried reaction solids and fly ash to the downstream particulate collector. Injecting a dry powdery sorbent, such as lime, into flue gas leaving the furnace/boiler effects acid gas removal, and the resulting dry reaction products are transported by water-unsaturated flue gas to the particulate collector. This latter process is universally called dry scrubbing, although humidification (quenching) of the flue gas may precede sorbent injection. In contrast, wet scrubbing produces a clean water-saturated flue gas whose temperature is somewhat lower than that of flue gas from semi-dry and dry scrubbing.

The simplest process for removing acid gases (HCl, HF, and SO_2) is dry alkaline sorbent injection into flue gas from the furnace/boiler followed by particulate collection. If totally dry sorbents are injected into the flue gas duct or reactor, they require substantial contact time with the target flue gas pollutant to achieve high removal. Thus a fabric filter is preferred when high acid gas removal is required because the filter cake provides a better sorbent/flue gas contact than an ESP. However, flue gas humidification may be needed to ensure that spark carryover from the furnace/boiler does not pose a hazard to the fabric bags. Figure 3 shows a simple dry sorbent injection control system with the option for sorbent injection into the flue gas duct or into a circulating fluid-bed reactor.

In Japan, where lime injection into the flue gas duct is the most prevalent HCl control method on MSW combustors [100 to 200 tonnes/day (110 to 220 tons/day)], the ESP is preferred for particulate collection. However, except in special areas, the particulate standard (see Table 1) and HCl standard are readily achieved by this control method. For example, an inlet HCl concentration ranging from 600 to 800 ppmv (corrected to 12% O_2) would require an HCl removal ranging from 28 to 46% to meet the HCl standard of 430 ppmv. Also, SO_2 control is generally not required as the uncontrolled SO_2 concentration at Japanese MSW incinerators is 60 to 100 ppmv and, except in special areas, the standard approximates 100 ppmv.

A pilot-scale study conducted for Environment Canada by Flakt-Canada indicated very effective acid gas control [>90% HCl and SO_2 removal, except for one test at 140°C (284°F)] when dry lime was injected into the reactor where the flue gas was humidified by water injection prior to contacting lime (see Figure 3) (6). Particulate collection was by a fabric filter. The pilot system evaluated is similar to systems in use in Western Europe (1).

Spray dryer (semi-dry or wet/dry) absorption is more effective than dry sorbent injection since its approach to saturation temperature is lower, and acid gas removal increases with decreasing approach to saturation temperature. Waste heat recovery may precede the spray dryer to improve energy recovery from the combustion gases while attaining the desired approach to saturation temperature. A spray dryer absorption system is shown in Figure 4.

Pilot unit and commercial MSW combustion unit tests have demonstrated HCl removals of 90% or more and SO_2 removals exceeding 70% with lime spray absorption followed by fabric filters. Similar results have been obtained with lime spray dryer absorber/ESP systems. However, the data were for short-term tests, and data are needed on the long-term performance of these acid gas control systems on commercial units.

Wet scrubbers, preceded by an ESP with 99% or greater control, often use sodium hydroxide (NaOH) in MSW combustion applications and are particularly favored where high trace heavy metals control is desired in addition to acid gas control. A heat exchanger downstream of the wet scrubber (Figure 5) can be used to condense water vapor from the water-saturated flue gas, and (through cooling of the flue gas) limit the visible steam plume leaving the stack as well as remove submicrometer (submicron) particles and metallic vapors. In removing trace metals from the liquid scrubber effluent, the effluent is treated to precipitate the metals. The remaining liquid effluent may also require pH treatment prior to its discharge to a nearby water stream or body, if permitted.

A wet scrubber preceded by a spray dryer/particulate collector is shown in Figure 6. This system has two acid gas venturi scrubbers in series, the first being a water prescrubber for HCl removal and the second using NaOH for SO_2 removal. The liquid effluents from both scrubbers (from the first venturi after neutralization by lime) are pumped to the spray dryer, which serves primarily to evaporate the liquids. The cooled flue gas leaves the spray dryer and passes through the fabric filter or ESP before entering the prescrubber. This system, now used in West Germany, is designed to operate with zero discharge of liquid effluent.

Wet scrubbers (Figure 5) are used on MSW combustion units in Japan and Western Europe. Both systems with wet scrubbers (Figures 5 and 6) can achieve 95% HCl removal, 90% SO_2 removal, and high removal of trace metals and organic compounds (dioxins/furans), and can be adapted to NO_x control. However, they are complex relative to dry and semi-dry scrubbers, the complexity increasing as the number of targeted pollutants increases.

Table 3 compares the effectiveness of acid gas control systems. It indicates that effective acid gas control can be attained with dry, semi-dry, or wet scrubbers. While HCl and HF control is relatively easy, SO_2 control is more difficult and is more effective with alkaline wet and semi-dry systems operating at and near flue gas saturation temperature, respectively.

Post Combustion NO_x Control

While NO_x control on MSW combustion units is not normally required by national standards (e.g., 250/300 ppmv in Japan and 500 ppmv in West Germany), local or state (or similar) regulations may require it. NO comprises 95% or more of the NO_x in flue gas, and because of its low reactivity and water solubility, NO is probably the most costly and difficult component of NO_x to remove. Wet scrubbing control of NO_x, recently introduced in Western Europe and proposed in Japan, is based on either oxidation-absorption or absorption-reduction. Selective catalytic reduction (SCR), originally developed in Japan for coal- and oil-fired plants, has also been applied to MSW combustion units.

The oxidation-absorption approach to NO_x control involves adding an oxidizer, such as sodium chlorite ($NaClO_2$), to the scrubber (Figure 5) or prescrubber (Figure 6) to convert NO in the flue gas to NO_2. For example, adding $NaClO_2$ to the first venturi scrubber (prescrubber) of Figure 6 and raising the pH to 3 or 4 result in NO's being oxidized to NO_2, which is then absorbed in this and the following SO_2 scrubber. The oxidation-absorption method is used in several West German plants (7).

Absorption-reduction to remove NO has been recently applied to MSW incinerators in West Germany (8). In this method, ferrous ions (Fe^{2+}) in the wet scrubber tie up NO in the liquid phase by forming ethylene diamine-tetracetic acid (EDTA) complexes. These complexes react with HSO_3^- and SO_3^{2-} ions (from SO_2 absorption) to yield N_2 and SO_4^{2-} as reaction products. NO_x control with this method is reported to be about $2 to 5 per tonne ($1.80 to 4.50 per ton) of waste processed below that of oxidation-absorption and is based on the cost of chemicals only (7).

Selective catalytic reduction for NO_x control is illustrated for conventional (high temperature) and low temperature catalysts in Figure 7. In conventional SCR, acid gas and heavy metals control must precede SCR for high NO_x removal and avoidance of early catalyst failure due to poisoning. If the thermal penalty of prescrubbing for acid gas and metals removal is acceptable, then 80 to 90% NO_x can be removed with a NH_3/NO molar ratio of 1.0 and about 5 ppmv NH_3 slip.

The SCR system with low temperature catalyst, also shown in Figure 7, is being used on three MSW combustion units in Japan (9) and has been selected for a proposed unit in California. The SCR unit follows the fabric filter of a dry lime injection system for HCl control, and the catalyst sees flue gas at 190 to 280°C (374 to 536°F) compared with 300 to 400°C (572 to 752°F) for conventional SCR. Performance tests in the Spring of 1987 on two new 65 tonne/day (72 ton/day) units served by one SCR unit gave 83% NO_x control (design value of 80%) with 140 ppmv NO_x at an inlet gas temperature of 200°C (392°F), an outlet NO_x value of 25 ppm, and a molar ratio of NH_3 to inlet flue gas NO_x of 1.05. Another test on a 150 tonne/day (165 ton/day) unit designed for 50% NO_x control indicated 57% NO_x removal when NO_x at 98 ppmv entered the SCR unit operating with flue gas at 246°C (475°F) and a molar ratio of 0.67. All NO_x concentration values given have been corrected to 12% O_2.

With a system similar to that in Figure 8, Mitsubishi claims: 98% HCl, 85% SO_x, and 80% NO_x removal with less than 20 ppm of each of these pollutants in the outlet flue gas; and 99% particulate (dust) removal with less than 20 mg/Nm^3 (0.008 gr/dscf) in the outlet flue gas (10).

Post-Combustion Organic Pollutant Control

The mechanisms of capturing dioxins, furans, and trace metals are not well understood. Their capture by condensation or attachment to particles is thought to be significant. Dioxin/furan attack and capture via caustic reagents and highly efficient particulate control is also suspected. Limited pilot plant data indicate that lime spray absorption followed by fabric filtration is highly effective for organic vapor control and superior to the spray absorber/ESP combination. In addition, lower flue gas temperature by spray absorption (11) and flue gas humidification before dry sorbent injection (6) favors improved organics removal.

Limited data are available on control system efficiencies for dioxins and furans. Only outlet concentrations are reported for tests performed to determine compliance of commercial units with their permit conditions. Several tests conducted recently for EPA in connection with compliance tests have collected both inlet and outlet organics data for lime spray absorption in combination with a fabric filter and an ESP, but the results are not yet available.

As noted earlier, the spray dryer absorber followed by fabric filtration was found to be more effective for organics control than when the spray-dried product was collected in an ESP. Table 4 summarizes these results, but the high and low temperatures noted were not given in Reference 11. It is believed that these temperatures were 150 and 110°C (302 and 230°F), respectively, based on subsequent information.

Heavy Metals Control

The removal of trace heavy metals from flue gas of MSW combustors is believed to be similar to that of organic pollutant control, where effective control of particles and low flue gas temperature are major factors. Sorbent reactions with metals, however, are not believed to be a major factor in metals control. The metals enter the air pollution control device as solids, liquids, and vapors, with some vapors becoming collectable solids and liquids as the flue gas cools. The flue gas temperature is usually decreased to 140°C (284°F) or less with lime spray absorption followed by particulate collection. With particulate removal by a fabric filter, 95% Hg removal was achieved with lime spray absorption, with and without solids recycle, compared with 94% Hg removal with dry lime injection (fluid-bed reactor) at the same fabric filter inlet gas temperature (140°C) (6). Other metals (arsenic, antimony, cadmium, chromium, lead, nickel, and zinc) were reduced by over 99% in both processes at the same temperature. While the lime spray absorption system was operated at only one temperature, the dry lime injection process was operated at lower [(110 and 125°C (230 and 257°F)] and higher temperatures [~209°C (408°F)] corresponding to no gas cooling by humidification. Very high metals control, with the exception of mercury, was obtained at the other temperatures, similar to that at 140°C. For some unexplained reason, the mercury vapor concentration data indicated a higher Hg concentration at the outlet (610 µg/Nm^3 at 8% O_2) than at the inlet (450 µg/Nm^3 at 8% O_2) for the >200°C (>392°F) test. Comparison of metals control in the Environment Canada (EC) tests with those of other dry scrubbing systems confirmed the EC test results and also suggested that flue gas cooling can improve mercury removal from flue gas.

Reported metals control data generally reveal 95% or higher control for most heavy metals except mercury. Vapor-phase mercury removal has been reported to be 75 to 85% with the lime spray absorption/fabric filter system and 35 to 45% with the lime spray absorption/ESP system (12). The spray absorber outlet temperature was 140°C (284°F) and the approach to saturation temperature was 85°C (153°F) in both scrubber systems.

With vapor condensation believed to be important to mercury control, wet scrubbers would appear to be better suited for this purpose than dry ones as they operate with flue gas saturation [~40°C (104°F)]. It also appears that $CaCl_2$, a product of calcium-based HCl scrubbers, is very active toward mercury and other metal vapors. Mercury collection methods are not well documented so that the choice of the most effective mercury control is still not uniform (13).

SUMMARY

Wet or dry scrubbers are effective for controlling pollutants (acid gases, trace organics, trace heavy metals, and particulate matter) produced in burning MSW waste. The choice of scrubber type depends on the pollutants to be controlled and the degree of control required. Dry sorbent (lime) injection with an ESP is used extensively in Japan for acid gas control, but wet scrubbing is preferred where high metals control is needed. The ESP/wet scrubber combination appears to be favored in West Germany for plants started up in the past decade and those expected to start up in the next several years. In the U. S., the lime spray absorption/fabric filter system is now frequently being selected for multi-pollutant control.

Acid gas removals of 90% or more have been achieved with a lime circulating fluid bed or lime spray dryer absorber preceding a fabric filter or ESP. Wet scrubbing preceded by an ESP is at least as effective as the systems noted when used to control acid gases. These systems are also effective for controlling organics and trace heavy metals, with mercury control appearing to be improved at lower temperatures and when a fabric filter rather than an ESP is used. Both the ESP and fabric filter can meet current particulate control requirements, but the fabric filter may have the edge for multi-pollutant control. More data, especially from commercial units, under long-term operation are needed to more fully quantify the performance of scrubbers designed to remove trace organic compounds and trace heavy metals.

LITERATURE CITED

1. Brna, T.G. and C.B. Sedman, "Waste Incineration and Emission Control Technologies," International Congress on Hazardous Materials Management, Chattanooga, TN (1987).

2. Seeker, W.R., W.S. Lanier, and M.P. Heap. Municipal Waste Combustion Study: Combustion Control of Organic Emissions, EPA/530-SW-087-021c (NTIS PB87-206090), (1987).

3. Hurst, B.E. and C.M. White, "Thermal DeNO$_x$: A Commercial Selective Non-Catalytic NO$_x$ Reduction Process for Waste to Energy Applications," ASME 12th Biennial National Waste Processing Conference, Dewey, CO (1986).

4. Overmoe, B.J. et al., "Influence of Coal Combustion on the Fate of Volatile and Char Nitrogen During Combustion," Nineteenth Symposium (Int.) on Combustion. The Combustion Institute, Pittsburgh, PA, p. 1271 (1982).

5. Memorandum: "Review and Update of PM Emissions Database for Solid Waste-fired Boilers," Radian Corporation, Research Triangle Park, NC (1986).

6. The National Incinerator Testing and Evaluation Program: Air Pollution Control Technology, Summary Report EPS 3/UP/2, Environment Canada, Ottawa, Ontario, Canada (1986).

7. Scrubber Absorber Newsletter, No. 152, The McIlvaine Company, Northbrook, IL (1987).

8. Brna, T.G., W. Ellison, and C. Jorgensen, "Cleaning of Municipal Waste Incinerator Flue Gas in Europe." ASME 13th National Waste Processing Conference and Exhibit, Philadelphia, PA (1988).

9. Mitsubishi SCR System for Municipal Refuse Incinerator: Measuring Results at Tokyo-Hikarigaoka and Iwatsuki, Mitsubishi Heavy Industries, Ltd., Yokohoma Dockyard and Machinery Works, Japan, April 14, 1987.

10. Mitsubishi Integrated Flue Gas Treatment System, Brochure HD10-07213, Mitsubishi Heavy Industries, Ltd., Tokyo, Japan.

11. Nielsen, K.K., J.T. Moeller, and S. Rasmussen, "Reduction of Dioxins and Furans by Spray Dryer Absorption from Incinerator Flue Gas," Dioxin 85, Bayreuth, West Germany (1985).

12. Moeller, J.T., C. Jorgensen, and B. Fallenkamp, "Dry Scrubbing of Toxic Incinerator Flue Gas by Spray Absorption," ENVITEC 83, Duesseldorf, West Germany (1983).

13. Scrubber Adsorber Newsletter, No. 145, The McIlvaine Company, Northbrook, IL (1986).

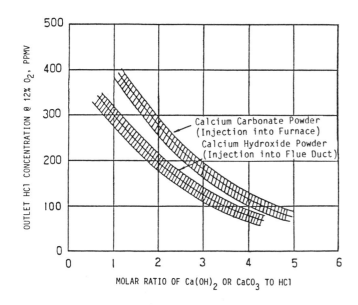

Figure 1. HCl removal by dry sorbent injection into furnace and flue gas duct (courtesy of Ishikawajima-Harima Heavy Industries Co., Ltd., Tokyo, Japan).

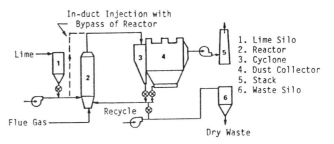

Figure 3. Dry sorbent flue gas cleaning by in-duct injection or a fluid-bed reactor (1).

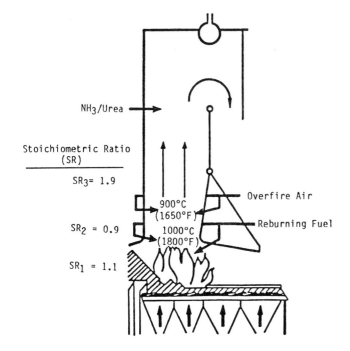

Figure 2. Reburning and thermal DeNO$_x$ in a MSW combustor (1).

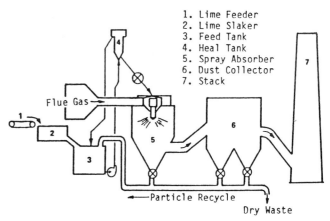

Figure 4. Spray dryer absorption (semi-dry or wet/dry) system (1).

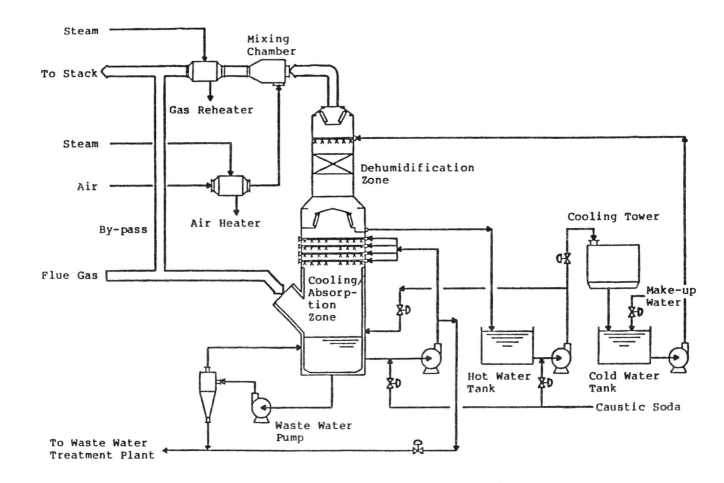

Figure 5. Wet flue gas scrubbing system (courtesy of Hitachi Zosen Corp., Osaka, Japan).

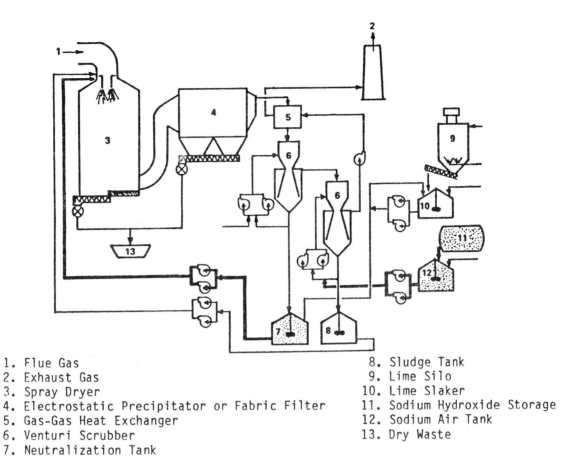

1. Flue Gas
2. Exhaust Gas
3. Spray Dryer
4. Electrostatic Precipitator or Fabric Filter
5. Gas-Gas Heat Exchanger
6. Venturi Scrubber
7. Neutralization Tank
8. Sludge Tank
9. Lime Silo
10. Lime Slaker
11. Sodium Hydroxide Storage
12. Sodium Air Tank
13. Dry Waste

Figure 6. Semi-dry/wet scrubbing system (1).

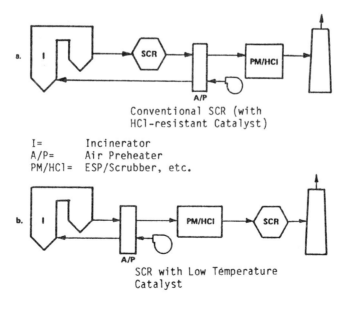

Figure 7. SCR options for MSW combustors (1).

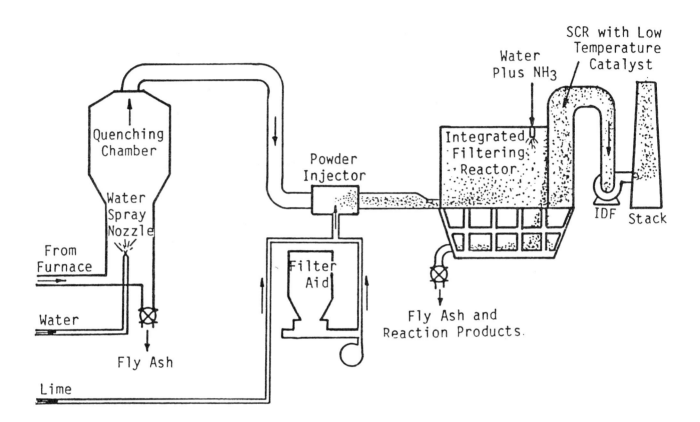

Figure 8. Integrated flue gas control system for waste-to-energy plant.

TABLE 1. SELECTED EMISSIONS STANDARDS FOR MUNICIPAL WASTE INCINERATORS (1)

	U.S.[a]	California	Connecticut	Michigan	Japan	Sweden[b]	West Germany
Solid Particulate Matter mg/m^3 (gr/dscf)	113 (0.046)	25[c] (0.01)	37 (0.015)	37 (0.015)	150[d] (0.061)	20 (0.008)	30 (0.012)
Carbon Monoxide, ppm	------	----------	----------	113(24-hr avg)	----------	------	80
Hydrogen Chloride	------	30 ppmv (scrubbers required)	90% reduction	90% reduction	430 ppmv (700 mg/m^3)	63 ppmv (100 ng/Nm^3)	31 ppmv (50 mg/Nm^3)
Sulfur Dioxide, ppm	------	30 [f]	135 (0.32 lb/10^6 Btu) [g]	86	Varies[e]	New SO_2 limits reduce all acids	35 (200 mg/Nm^3)
Dioxins Measured as 2,3,7,8 - tetra-chlorodibenzo-p-dioxins (TCDD)	------	----------	----------	--------------	----------	Existing plants: 0.5-2.0 ng/Nm^3 New plants: 0.1 ng/Nm^3	------------
Total Organics, mg/m^3	------	----------	----------	--------------	----------	--------	20
Mercury+Cadmium+Thallium mg/m^3 (includes vapors)	------	----------	----------	--------------	----------	0.08 (Hg only)	0.2
Gas Correction	12% CO_2 dry	12% CO_2 dry	12% CO_2 dry	12% CO_2 dry	12% O_2 dry	10% CO_2 dry	11% O_2 dry

[a] Revised NSPS for pollutants scheduled to be proposed in 1989.
[b] Swedish Environmental Protection Board's "Temporary Emission Goals," July 1986.
[c] California regulations permit more stringent local limits. Two state guidelines are reported: 25 mg/m^3 (0.01 gr/dscf) for total solid particulates (TSP) and 20 mg/m^3 (0.008 gr/dscf) for particles less than 2 μm.
[d] Based on continuous gas flows >40,000 m^3/h (>25,280 scfm). For flows ≤25,280 scfm, the particulate matter standard is 500 mg/m^3 (0.20 gr/dscf). For new plants in special areas, this standard is 80 mg/m^3 (0.033 gr/dscf) for plants having >40,000 m^3/h and 150 mg/m^3 for those having ≤40,000 m^3/h.
[e] Based on formula related to stack height and plant location. Typically, plant sulfur dioxide emissions range from 60 to 100 ppm so that control is not required except for new plants in special areas.
[f] Pollutant control requires use of the Best Available Control Technology (BACT), but no technology is specified.
[g] The use of dry gas scrubbers and baghouses is expected to improve removal over ESPs alone.

TABLE 2. NO_x CONTROL OPTIONS FOR MSW PLANTS (COURTESY OF HITACHI ZOSEN CORP., OSAKA, JAPAN)

Systems	Attainable Control Level (ppm v/v, dry, 12% O_2)	Remarks
Uncontrolled	150–130	
Automatic Combustion Control	120–100	
Water Spray in Combustion Chamber	100–80	used with automatic combustion control
Flue Gas Recirculation	80–60	ditto
Ammonia Injection without Catalyst	70–50	ditto
Ammonia Injection with Catalyst	60–40	
Wet Scrubber with Oxidization	60–40	

Note: NO_x concentration is corrected by O_2 percentage in Japan, not by CO_2 percentage.

TABLE 3. EFFECTIVENESS OF ACID GAS CONTROLS (% REMOVAL)

	Pollutant		
Control System	HCl	HF	SO_2
Dry Injection + Fabric Filter (FF)[a]	80	98	50
Dry Injection + Entrained Fluid Bed Reactor + ESP[b]	90	99	60
Spray Dryer Absorber + ESP[c]	95+	99	50-70
(Recycle)[c]	(95+)	(99)	(70-90)
Spray Dryer Absorber + Fabric Filter[c]	95+	99	70-90
(Recycle)[c]	(95+)	(99)	(80-95)
Spray Dryer Absorber + Dry Injection + ESP or FF[d]	95+	99	90+
ESP + Wet Scrubber[e]	95+	99	90+
Spray Dryer + Wet Scrubber(s) + ESP or FF[e]	95+	99	90+

[a] T = 160-180°C (320-356°F)
[b] T = 230°C (446°F)
[c] T = 140°-160°C (284-320°F)
[d] T = 200°C (392°F)
[e] T = 40-50°C (104-122°F)
   T = the temperature at the exit of the control device.

TABLE 4. SPRAY DRYER CONTROL OF SELECTED ORGANIC POLLUTANTS (11)

Control System (% Removal)

Compound	Full Scale SD + ESP	Pilot Plant	
		SD + FF 150°C (302°F)	SD + FF 110°C (230°F)
Dioxins			
tetra CDD	48	< 52	> 97
penta CDD	51	75	> 99.6
hexa CDD	73	93	> 99.5
hepta CDD	83	82	> 99.6
octa CDD	89	NA	> 99.8
Furans			
tetra CDF	65	98	> 99.4
penta CDF	64	88	> 99.6
hexa CDF	82	86	> 99.7
hepta CDF	83	92	> 99.8
octa CDF	85	NA	> 99.8

NOTES:

SD= spray dryer absorber
ESP= electrostatic precipitator
FF= fabric filter
Temperatures shown are flue gas values at the spray absorber outlet. They were not given in Reference 11 and are based on subsequent information.

REGULATORY ANALYSIS OF POLLUTANT EMISSIONS, INCLUDING POLYCHLORINATED DIBENZO-P-DIOXINS (CDDs) AND DIBENZOFURANS (CDFs), FROM THE STACKS OF MUNICIPAL WASTE COMBUSTORS

David H. Cleverly, Rayburn M. Morrison,
Brenda L. Riddle and Robert G. Kellam ■ U.S. Environmental Protection Agency, Office of Air Quality Planning and Standards, Research Triangle Park, NC 27711

Municipal waste combustors (MWCs) are known to emit into the air quantities of CDDs, CDFs and other carcinogenic and toxic organic and inorganic compounds from the stack. In support of a regulatory determination, an impact analysis, including a quantitative risk assessment of 13 pollutants in the air emissions, was conducted to evaluate the impacts of further controls on emissions. This paper describes the Environmental Protection Agency's (EPA) regulatory analysis of dioxins and other pollutant emissions from MWCs, and discusses the Agency's findings.

INTRODUCTION AND BACKGROUND

The combustion of municipal solid waste (MSW) represents an increasingly important element of the solid waste disposal problem in the United States. At the present time, approximately 111 MWCs incinerate in excess of 6×10^6 metric tons of MSW annually. This amount represents only 5 to 6 percent of the total annual mass of MSW, however the U.S. EPA estimates that, by the mid-1990's, combustion could account for as much as 25 percent of domestic MSW disposal in as many as 300 facilities.

The anticipated growth in the industry is largely the result of a consistent growth in the volume of MSW and increasing concern about the continuous availability and environmental impacts of the disposal of MSW in landfills. While 80 percent of MSW is currently disposed of in about 10,000 landfills nationwide, many of these facilities are nearing design capacity, and the siting of new landfills has become increasingly difficult, particularly in densely populated areas.

Although the combustion of MSW does not eliminate the need for landfilling, since residual ash must be disposed of, it does reduce the volume of waste 70 to 90 percent thus extending the operating lifetime of landfills. Another benefit of incineration is the ability to recover energy from the heat of combustion in the form of steam or hot water which can be used to partially offset the energy requirements of the facility or sold to industrial clients or the local utility net.

Disposal of MSW by combustion, however, releases potentially harmful pollutants into the atmosphere. Currently EPA limits the emission of particulate matter (PM) to 180 milligrams per dry standard cubic meter (mg/dscm) (corrected to 12% CO_2) of stack gas at MWCs built before 1986. In 1986 EPA promulgated a PM standard equivalent to the emission of 70 mg/dscm for industrial boilers including new energy recovery MWCs with capacities greater than approximately 180 metric tons/day. The present control of emissions is not intended to address health concerns specific to individual chemical constituents or other pollutants of MWC emissions. Because the current standard for the control of PM emissions may not be sufficiently health protective, analysis related to the need for additional Federal regulation was undertaken. The following sections describe the methods and results of EPA's most current analysis of MWC emissions, and EPA's recent determination that further regulation of these emissions is warranted.

SOURCES AND EMISSIONS

The MWC industry in the U.S. can

be classified into three principal design types: massburn, modular, and refuse derived fuel (RDF). The existing stock of MWCs can be further classified as having or not having heat recovery boilers. The massburn and modular facilities usually combust the waste as received from refuse trucks without substantial recycling, preprocessing, or source-separating the waste. Mass burn facilities range in size from 45 to 2,720 metric tons per day of combustion capacity. Modular facilities are typically 4.5 to 91 metric tons/day in capacity. The third major class of MWCs incinerates RDF, which is a fuel that has been prepared by shredding, sorting, separating and trommeling MSW to reduce the noncombustible content. These facilities range in capacity from 227 to 2,720 metric tons per day. All RDF MWCs have boilers to recover heat. While some RDF units are designed to be co-fired with fossil fuels, only facilities burning RDF were defined as MWCs in this study. The number and total design capacity of existing MWC facilities is summarized in Table 1.

Table 1. Existing And Projected MWCs In The U.S.

DESIGN TYPE	EXISTING MWCs		PROJECTED MWCs	
	NO. MWCs	CAPACITY (MT/day)	NO. MWCs	CAPACITY (MT/day)
A. MASSBURN				
- No heat recovery	21	13,000	0	0
- With heat recovery	24	20,100	118	113,000
B. MODULAR				
- No heat recovery	17	600	0	0
- With heat recovery	39	3,900	24	5,000
C. RDF				
- With heat recovery	10	11,400	31	39,000
D. UNKNOWN	0	0	37	36,000
E. TOTAL	111	49,000	210	193,000

The EPA has estimated the projected growth of MWCs from 1986 to 1995 (EPA, 1987a) in terms of number of facilities and total capacity. These data are summarized in Table 1. Projected facilities are defined as planned, but not yet operating, and are either under construction, have contracts for construction, or have been formally proposed.

The evaluation of stack emissions from MWCs was limited to pollutants for which emission test data and some indication of public health or welfare concern were available. Data were sufficient for analysis of 13 pollutants or classes of pollutants as summarized in Table 2. On a total mass basis, the predominant emissions are carbon monoxide, hydrogen chloride, nitrogen oxides, and sulfur oxides.

Table 2. Current and Projected Emissions From MWCs (MT/Yr)

	EXISTING MWC[a]	PROJECTED MWC[b]
Arsenic	2.7	13.0
Beryllium	0.1	0.6
Cadmium	10.4	19.9
Carbon monoxide	20,000.0	84,500.0
Chlorobenzenes	3.8	2.5
Chlorophenols	5.4	3.5
Chromium^{+6}	0.6	2.6
TCDD toxic equiv.	0.024	0.021
Formaldehyde	58.6	119.0
Hydrogen chloride	47,700.0	195,000.0
Lead	341.0	1260.0
Mercury	68.0	148.0
Nitrogen oxides	30,500.0	134,000.0
Particulate matter	10,400.0	26,700.0
PCB	5.0	21.0
PAH	1.0	4.0
Sulfur oxides	13,000.0	63,000.0

[a] Based on existing air pollution control.
[b] Based on new MWCs achieving 99% control of PM.

EMISSIONS OF CDDs/CDFs

The emissions of CDDs/CDFs from MWCs are a phenomena well documented by stack sampling, but the precise mechanistic explanation remains unknown. Various theories have been proposed to explain this occurrence to include: (a) emissions are a result of CDDs/CDFs occurring in the bleached paper or other MSW constituents, and a portion survives the incineration process; (b) emissions are a result of the de novo synthesis from precursors such as PCB, pentachlorophenols, and chlorinated benzenes; (c) emissions are a result of synthesis from materials not chemically related to CDDs/CDFs such as petroleum products, chlorinated hydrocarbons, inorganic chloride ions and plastics (Hutzinger, 1985). The magnitude and rate of emission is thought to reflect formation downstream of the furnace in regions of the incinerator where the combustion offgases have cooled to 250° - 450°C, and that formation occurs on active sites on surfaces of particulate matter (Commoner, 1987; Vogg, 1987). Most of the congeners of the 75 CDDs and 135 CDFs have been detected in MWC emissions (US EPA, 1987e). Since refuse is a heterogenous waste, it is not unlikely that most, if not all of the formation mechanisms suggested may be operative during combustion.

The rate of emission can be discussed in terms of the toxic equivalent of the emissions of various congeners to 2,3,7,8-tetrachlorodibenzo-p-dioxin (EPA, 1987b). The EPA estimates that 24 kg of TCDD toxic equivalence (TEQ) are emitted annually from existing MWCs in the U.S. In the absence of additional Federal regulation, emissions from the 210 projected new facilities could add an additional 21 kg TCDD TEQ/yr, for a total of 45 kg TCDD TEQ/yr by 1995. Estimates can only be approximate since actual emissions of CDDs/CDFs are known to vary by as much as three orders of magnitude across the population of listed individual incinerators.

For the purposes of exposure and risk analysis, a range of TCDD TEQ emissions from MWCS have been incorporated in the analyses as shown in Table 3. This range accounts for the uncertainty in the distribution of toxic isomers within homologue groups, as well as uncertainty in the measurement of trace quantities of CDDs/CDFs in the stack gases with conventional sampling protocols (EPA, 1987c). In deriving emission factors, a distinction was made between existing and projected MWCs. This distinction was based on the premise that MWCs currently marketed represent distinct improvements in design, combustion efficiency, and pollution control when compared to facilities planned and built just a decade ago. The CDD/CDF emission factors were calculated by averaging emissions for MWCs representing subcategories of technologies, and dividing average emissions by the rate of MSW fed to the incinerator (e.g., grams nation of the magnitude of population exposure. In the absence of monitored of TCDD TEQ per megagram of MSW burned). Emission factors for each MWC technology were determined by calculating the average emission rate for all facilities of a given type, e.g., massburn, modular and RDF.

Table 3. Range Of TCDD Equivalent Emissions from MWC (Milligrams/Metric Ton)

FACILITY TYPE	EMISSION FACTOR	
	EXISTING	PROJECTED
Massburn - Heat Recovery	0.14 - 3.2	0.008 - 0.2
Massburn - Nonheat Recovery	0.44 - 10.3	N/A
Refuse Derived Fuel (RDF)	0.3 - 6.7	0.2 - 4.0
Modular	0.01 - 0.2	0.01 - 0.2

EXPOSURE ASSESSMENT

Estimation of the potential human health risks associated with pollutants emitted from the stacks of MWCs requires estimation of the concentrations of specific constituents to which the population may be exposed, and determiambient data, the EPA used mathematical models to predict the atmospheric dispersion of emissions and subsequent potential for human exposure. This is described in detail in an available EPA report (EPA, 1987c). Estimates of population exposure to ambient air concentrations of the emitted pollutants were developed using the EPA's Human Exposure Model (HEM) (EPA, 1986a). The HEM accepts as inputs the locations and emissions characteristic of actual or representative sources. This information is combined with population census and meteorological data to estimate the magnitude and distribution of population exposure through dispersion of the pollutant emissions. For the existing MWCs, an exposure analysis of each facility operating in the U.S. was conducted. Projected MWCs were represented by model plants developed from information from State permits describing size, capacity and stack parameters of planned units. Population exposure was extrapolated to the nation from the model plants by the projected national throughput for each technology subcategory.

For both existing and projected MWCs, assumptions were made regarding baseline control and stricter Federal control of pollutant emissions. These control scenarios are summarized in Table 4. For the existing and projected MWCs the baseline scenario assumed uniform use of electrostatic precipitators (ESPs) as the particulate control device. Based on available test data, ESPs were assumed to be 20 percent effective in the capture of organic pollutants (EPA, 1987c). The additional control scenario assumed the uniform application of dry alkaline scrubbers combined with fabric filters. Under this scenario, organic pollutants capture improved to a 95 percent collection efficiency based on actual test data (Environment Canada, 1986; EPA, 1987d). Metal emissions were assumed to be proportional to particulate emissions, except mercury which predominates in vapor phase at temperatures above 200°C.

NATIONAL ESTIMATES OF CANCER RISK

For MWC emissions a major health con-

Table 4. Control Scenarios Modeled For Existing and Projected Municipal Waste Combustors

POPULATION/ CONTROL SCENARIO	TECHNOLOGY	Control Efficiency (%)			
		ORGANICS	Metals (except Hg)	Hg	HCl
Existing					
Baseline	ESP	20	*	30	0
Controlled	Dry alkaline scrubber/ fabric filter	95	99.5	50	90
Projected					
Baseline	ESP	20	99	30	0
Controlled	Dry alkaline scrubber/ fabric filter	95	99.5	50	90

*Proportional to actual PM control levels.

cern is cancer. The estimates of population exposure to the pollutants were combined with a measure of carcinogenic potency for each chemical in the emissions to predict a probability of cancer. In the case of 2,3,7,8-TCDD, the carcinogenic data observed in a study by Kociba et al., (1979) was applied to the modified multistage model developed by Crump and Watson (1979) to estimate the carcinogenic potency. The 95 percent upper-limit carcinogenic potency for 2,3,7,8TCDD was determined for humans to be 1.56×10^5 (mg/kg/day)$^{-1}$ (US EPA, 1985). The cancer risk for continuous lifetime 70-year exposure to an ambient air concentration of 1 picogram m^{-3} of air of 2,3,7,8-TCDD has been estimated to be 3.3×10^{-5} (EPA, 1985), by assuming a breathing rate of 20 m^3 day^{-1} for a 70 kg man, and 75 percent of the inhaled material will be absorbed. Two measures of cancer risk are estimated: the aggregate cancer risk expressed as excess annual cancer incidence (expected cases/year), and the maximum individual lifetime risk (MIR) expressed as a lifetime probability of cancer for the most exposed populations. The MIR is often expressed in scientific notation as a negative exponent of 10, whereby a risk of 1 chance in 10,000 is expressed as 10^{-4}; a risk of one in one million as 10^{-6}, etc.

The ranges in the estimated cancer risks resulting from inhalation exposure to predicted ambient concentrations of MWC emissions are depicted in Table 5. These ranges of risk result from the quantitative uncertainties inherent in expressing carciogenic risk for the individual compounds as well as classes of similar compounds of concern (EPA, 1987c). Other risk assessment uncertainties that may yield either over or under estimates of risk include: the assumption of additivity of risk across carcinogens; the limiting of carcinogenic potency estimation to discrete compounds in the emissions; and the modeling rather than actual measurement of population exposure. Given these uncertainties, the analysis suggests that most of the estimated cancer risk associated with direct inhalation exposure to MWC emissions is attributable to the class of CDD/CDF, measured as the toxic equivalent to 2,3,7,8-TCDD. Thus, 2 to 40 and 0.8 to 20 cancer cases per year may result from baseline CDD/CDF emissions from existing and projected MWCs, respectively. The MIR ranges from one chance of cancer per one million (1×10^{-6}) to one chance per one thousand (1×10^{-3}) for existing MWCs, and one chance in one million (1×10^{-6}) to one chance in ten thousand (1×10^{-4}) for projected MWCs. Thus, in the absence of further control of emissions, CDD/CDF from all 321 existing and projected MWCs would be estimated to cause about 4 to 60 cancer cases per year in the exposed U.S. population.

There remain basic questions concerning the mechanism of carcinogenesis of dioxins. The models used by the US EPA implicitly assume that dioxin acts as an initiator of cancer. If, however, dioxin acts exclusively as a promoting agent to amplify the carcinogenic response of other direct acting carcinogens, then the present model may not be entirely appropriate. The chronic exposure studies of 2,3,7,8-TCDD indicate initiating events of organs other than the liver and skin, whereas subchronic bioassays indicate a promoting action in liver and skin carcinogenesis in animals (Mukerjee and Cleverly, 1987). In addition equivocal epidemiologic evidence suggests an association with soft tissue sarcoma and non-Hodgkins's lymphoma (Mukerjee and Cleverly, 1987). A change in the assumption on which cancer potency is based could possibly lead to a reduction in the current estimates of CDD/CDF cancer risks. However, these estimates did not consider the longevity of the compounds in the environment, their absorption into human adipose tissues and breast milk, nor their half-life in humans now estimated to be greater than seven years (Pirkle, 1987).

The EPA has also begun preliminary analysis of the potential for exposure from the deposition of emitted CDDs/CDFs, and subsequent human contact through indirect exposure pathways (EPA, 1986b, 1987c). The preliminary analysis indicates that indirect exposure via dioxin contaminated foods resulting from MWC emissions deposited over

Table 5. Ranges In The Contribution Of MWC Emission Constituents To Estimated Annual Cancer Incidence And Maximum Individual Lifetime Cancer Risk

Pollutant	Existing MWC Annual Cancer Incidence[a,d]	Existing MWC Maximum Individual Risk Range[b,e]	Projected MWC Annual Cancer Incidence[a,d]	Projected MWC Maximum Individual Risk Range[b,e]
Chlorinated dioxins and dibenzofurans (CDD/CDF)	2 to 40	10^{-6} to 10^{-3}	0.8 to 20	10^{-6} to 10^{-4}
Chlorophenols	0.0001 to 0.0003	10^{-9} to 10^{-8}	0.0001 to 0.0003	10^{-10} to 10^{-9}
Chlorobenzenes	0.009 to 0.02	10^{-7} to 10^{-6}	0.004 to 0.01	10^{-9} to 10^{-7}
Formaldehyde	0.009	10^{-8}	0.02	10^{-8} to 10^{-7}
Polycylic aromatic hydrocarbons	0.01 to 0.6	10^{-7} to 10^{-5}	0.05 to 3.0	10^{-7} to 10^{-5}
Polychlorinated biphenyls	0.02	10^{-8} to 10^{-5}	0.2	10^{-9} to 10^{-6}
Arsenic	0.2	10^{-7} to 10^{-4}	0.1	10^{-8} to 10^{-7}
Beryllium	0.02	10^{-9} to 10^{-6}	0.001	10^{-11} to 10^{-8}
Cadmium	0.2	10^{-6} to 10^{-4}	0.2	10^{-7} to 10^{-6}
Chromium^{+6}	0.2	10^{-7} to 10^{-4}	0.1	10^{-7} to 10^{-6}
Rounded Total:[c]	2 to 40	10^{-6} to 10^{-3}	2 to 20	10^{-6} to 10^{-4}

[a] The ranges in annual cancer incidence reflect assumptions made regarding the potential carcinogenicity of classes of organic compounds.
[b] The ranges in maximum individual lifetime cancer risk reflect differences in emissions and the evaluation of emissions from MWC technologies within the existing and proposed categories.
[c] Apparent errors in totals are due to intentional rounding to one significant figure.
[d] Annual cancer incidence is defined as the average number of excess cancer cases expected annually in the exposed population.
[e] Maximum individual risk is defined as the probability of contracting cancer following a lifetime exposure at the maximum modeled long-term ambient concentration. The probability is expressed as a negative exponent of 10. A risk of 1 chance in 10,000 is expressed as 10^{-4}.

30 years may be comparable to exposures through direct inhalation. This observation warrants further study.

REGULATORY DETERMINATION

Based on the assessment, the US EPA Administrator has judged that MWC emissions can reasonably be anticipated to contribute to the endangerment of the public health and welfare (Federal Register 1987), and warrant further regulation under the Clean Air Act (CAA). The EPA intends to revise the new source performance standard (NSPS) for control of MWC emissions under section 111(b) of the CAA. The NSPS will reflect the best system of continuous emission reduction that has been demonstrated for MWCs. The EPA has also announced an intent to designate one or more unregulated pollutants in MWC emissions. The designation will require the issuance of existing source guidelines (under section 111(d) of the CAA) for use by the States in the development of emission standards for existing MWCs. The standards will include emission limits and quantitative requirements for monitoring. The proposal of the NSPS and draft existing source guidelines is currently scheduled for November 1989. The NSPS under development will apply to MWCs for which construction has commenced after proposal. Promulgation of the final NSPS requirements are scheduled for December 1990.

In the interim, the EPA has issued operational guidance for use by State and local agencies with permitting authority. This operational guidance is intended to be followed by these regulatory agencies in reviewing best available control technology (BACT) determinations for new or modified MWCs. Under this BACT guidance, dry alkaline scrubbers followed by electrostatic precipitators or fabric filters are deemed effective in controlling toxic organic pollutants (such as CDDs/CDFs), metal emissions, and acid gases.

CONCLUSION

The EPA has evaluated pollutant emissions from MWC currently operating as well as expected to begin operation in the next several years. From these data a nationwide cancer risk assessment was undertaken to determine the need for further control of stack emissions. The analysis suggests that by the mid-1990s, if no additional control were to occur, it is possible that 4 to 60 cancer cases per year would occur in the exposed US population. Most of the carcinogenic risk is attributable to the emission of highly toxic CDDs and CDFs. The analysis was an important factor in EPA's determination in July 1987 that further control of the release of pollutants to the air from the stacks of MWCs was warranted. Uniform application of dry alkaline scrubbers combined with either fabric filters or ESPs, as pollution control devices, could potentially reduce the carcinogenic risk associated with the emission of CDDs/CDFs and other pollutants by over one order of magnitude.

REFERENCES

Commoner, B. et al., 1987. The origin and health risks of PCDD and PCDF. Waste Management and Research, 5:327-346.

Crump, K.S., and Watson, W.W., 1979. Global 79: A Fortran Program to Extrapolate Dichotomous Animal Carcinogenicity Data to Low Dose. NIEHS Contract #I-ES-2123.

Environment Canada, 1986. The National Incinerator Testing and Evaluation Program: Air Pollution Control Technology. Report EPA 3/UP/2, September.

Federal Register, 1987. U.S. Environmental Protection Agency: Assessment of Municipal Waste Combustor Emissions under the Clean Air Act, 52:25399-25408, Tuesday, July 7.

Kociba, R.J., et al., 1979. Long-term toxicological studies of 2,3,7,8-tetrachlorodibenzo-p-dioxin (TCDD) in laboratory animals. Annals of the New York Academy of Science, 320:397-404.

Hutzinger, O. et al., 1985. Sources and fate of PCDDs and PCDFs: An Overview. Chemosphere, 14:581-600.

Mukerjee, D. and D. Cleverly, 1987. Risks from exposure to polychlorinated dibenzo-p-dioxins and dibenzofurans emitted from municipal incinerators. Waste Management and Research, 5:269-283.

Pirkle, J. et al., 1987. Estimates of the half-life of 2,3,7,8-tetrachlorodibenzo-p-dioxin in ranch hand veterans. Centers for Disease Control. Presented at 7th International Symposia on Chlorinated Dioxins and Related Compounds, October 4-9, 1987, University of Nevada, Las Vegas.

U.S. Environmental Protection Agency, 1987a. Characterization of the Municipal Waste Combustion Industry, EPA/530-9W-87-021h, June.

U.S. Environmental Protection Agency, 1987b. Interim Procedures for Estimating Risks Associated with Exposures to Mixtures of Chlorinated Dibenzo-p-dioxins and Dibenzo-furans (CDDs and CDFs), EPA/625/3-87/012, March.

U.S. Environmental Protection Agency, 1987c. Assessment of Health Risks Associated with Municipal Waste Combustion Emissions, EPA/530-9W-87-021g, September.

U.S. Environmental Protection Agency, 1987d. Municipal Waste Combustion Multipollutant Study: Emission Test Report, Marion County Solid Waste-to-Energy Facility. Volume 1,2: Summary of Results. EMB Report #86-MIN-03. September.

U.S. Environmental Protection Agency, 1987e. Biological Marker Study for Sources of Polychlorinated Dibenzo-p-dioxins and Dibenzofurans in the Environment. Draft Draft final report prepared by Pacific Environmental Services. EPA 450/5-87-007. September.

U.S. Environmental Protection Agency, 1986a. User's Manual for the Human Exposure Model (HEM), EPA-450/5-86-001, June.

U.S. Environmental Protection Agency, 1986b. Methodology for the Assessment of Health Risks Associated with Multiple Pathway Exposure to Municipal Waste Combustor Emissions. For Review to EPA's Science Advisory Board, October.

U.S. Environmental Protection Agency, 1985. Health Assessment Document for Polychlorinated Dibenzo-p-dioxins. EPA/600/8-84/014F. Final Report. September.

Vogg, H. et al., 1987. Recent findings on the formation and decomposition of PCDD/PCDF in municipal solid waste incineration. Waste Management and Research, 5:285-294.

THE HAZARDS OF MUNICIPAL INCINERATOR ASH AND FUNDAMENTAL OBJECTIVES OF ASH MANAGEMENT

Richard A. Denison ■ Toxic Chemicals Program, Environmental Defense Fund, 1616 P Street, NW Suite 150, Washington, DC 20036

Recent data are discussed that corroborate earlier indications that municipal solid waste (MSW) incinerator ash is hazardous. These data demonstrate that: 1) ash contains high levels of several highly toxic metals, and can also contain dangerous levels of dioxins; 2) certain of the metals—lead and cadmium, in particular—are readily leachable from ash at levels that frequently exceed the limits defining a hazardous waste; 3) incineration concentrates and mobilizes the metals present in MSW, and can create dioxins, opening up several new pathways of exposure to these toxins; and 4) ash is toxic when tested by several means in addition to the EP toxicity test. Several approaches to ensuring proper ash management are discussed in light of these findings.

INTRODUCTION

Foremost among the underexplored risks of MSW incineration are the hazards posed by the frequent presence of high levels of dioxins and the routine presence of several toxic metals in ash residues. Ironically, the growing use of more efficient air pollution control devices on modern incinerators results in ash containing even higher levels of these toxic substances in even more bioavailable forms.

The most fundamental and important measure of ash toxicity is its <u>total</u> toxic metal and dioxin content, given the potential for direct exposure (e.g., inhalation and adsorption into the lung or direct ingestion of toxin-laden ash particles). Indeed, a full accounting of the hazards of ash posed during all phases of its management requires knowledge of its total chemical composition.

Table I compares typical concentration ranges of lead and cadmium in MSW incinerator fly ash to those found in natural soils, illustrating their extreme enrichment in this waste. Less extreme but still significant metal enrichment characterizes bottom ash. Table I also illustrates that the total metal content of incinerator ash is comparable to other materials clearly regarded or classified as hazardous. Emission control sludge from secondary lead smelters -- a waste listed as hazardous under federal regulations -- exhibits a range of lead and cadmium content quite similar to incinerator fly ash ([1]). A recent Washington state study ([2]) documented levels of several carcinogenic metals in both fly and bottom ash that were sufficiently high to classify the ashes as dangerous or extremely hazardous wastes under state regulations.

Metals are chemical elements, and can neither be created nor destroyed by incineration; their amounts in the waste stream before incineration must therefore equal the sum of their amounts in air emissions and ash left after incineration. The process of incineration, however, is uniquely unsuited for managing metals. Incineration essentially destroys the bulky matrix -- paper, plastics, or other materials

Table I. Metals Concentrations in Incinerator Fly Ash, Secondary Lead Smelter Sludge, and Natural Soils

RANGE OF CONCENTRATIONS
(parts per million)

METAL	FLY ASH	SMELTER SLUDGE[a]	NATURAL SOILS[b]
Lead	2,300-50,000	up to 50,000	10-13
Cadmium	100-2,000	340	0.1-0.2

a EPA, 1980 (Reference 1)
b Vogg et al., 1986 (Reference 39)

-- which contains metals in MSW and which acts to retard their entrance and dispersion into the environment (see, for example, ref. 3). In this respect, incinerators can be compared to secondary metal smelters; by burning combustible materials they release metals, which are subsequently mobilized in air emissions or concentrated in the residues in highly bioavailable form.

In addition to metals, highly toxic dioxins have been detected in all samples of incinerator fly ash tested, in some cases at levels that greatly exceed government guidelines (4, Denison, manuscript in preparation). While dioxins appear to be lower in fly ash from newer facilities, the ability of such facilities to consistently achieve acceptably low levels remains to be demonstrated. As is the case for metals, more efficient air pollution control devices will act to increase the concentration of dioxins detected in ash residues. In addition, combustion controls designed to increase burnout and reduce dioxin formation may also increase ash toxicity. Recent data indicate that the bioavailability of fly ash-bound dioxins to fish is inversely related to the organic carbon content of the ash (5), so that reduction in dioxin air emissions through better combustion and stack controls may nevertheless yield an ash that poses a greater risk of dioxin exposure.

LEACHABILITY OF METALS IN INCINERATOR ASH

The leachability of metals present in incinerator ash is another measure of hazard. Table II presents a summary of data on ash from more than 30 U.S. incinerators tested for leaching using the federally mandated Extraction Procedure (or EP; 40 C.F.R. 261.24). These test data -- on ash from new and old facilities employing a wide range of technologies -- demonstrate that:

* virtually every sample of fly ash tested exceeds federal standards defining a hazardous waste, usually for both lead and cadmium.

* half of the combined fly and bottom ash samples tested also exceed the standards, typically for lead.

These results indicate that incinerator ash routinely exhibits the EP toxicity characteristic of hazardous waste. In the heat of the debate surrounding the suitability of the EP test for incinerator ash, one fact is frequently overlooked: the vast majority of this ash is disposed of in sanitary landfills along with unburned waste (4) -- exactly the disposal scenario which the EP test is designed to simulate.

The increased leachability of metals in incinerator ash arises from several phenomena associated with combustion. First, several toxic metals are volatilized and then condense onto the surface of fly ash particles, and their concentrations increase with decreasing particle size (6-9). The small particle size increases the available surface area exposed to the leaching medium (6), and the presence of metals at or near the surface of such particles also enhances leachability (7,10). In addition, the high chlorine content of MSW results in significant complexation of metals as metal chlorides (11-13), which generally are much more soluble in water than most other speciated forms of metals.

Another chemical property of certain metals becomes critical when evaluating the quality of ash generated by facilities equipped with acid gas scrubbers. Through the operation of these devices, a slurry or powder of lime is introduced to neutralize acid gases, and is intimately mixed into fly

Table II. Summary of Available Extraction Procedure Toxicity Test Data for Lead and Cadmium from MSW Incinerator Ash

	LEAD	CADMIUM	EITHER
FLY ASH (19 Facilities)			
No. of Samples Analyzed	87	85	87
No. of Samples Over EP Limit	83	83	87
% of Samples Over EP Limit	95%	98%	100%
Average of All Samples (mg/L)	$\underline{23.0}^a$	$\underline{28.4}$	--
BOTTOM ASH (10 Facilities)			
No. of Samples Analyzed	245	210	245
No. of Samples Over EP Limit	93	4	94
% of Samples Over EP Limit	38%	2%	38%
Average of All Samples (mg/L)	$\underline{6.7}$	0.19	--
COMBINED ASH (26 Facilities)			
No. of Samples Analyzed	366	272	366
No. of Samples Over EP Limit	171	54	176
% of Samples Over EP Limit	47%	20%	48%
Average of All Samples (mg/L)	$\underline{7.6}$	0.68	--

a Underlined values exceed EP limits defining a hazardous waste: lead: 5.0 milligrams per liter (mg/L) cadmium: 1.0 mg/L ash to form a scrubber residue removed by downstream particulate control devices. For the several U.S. facilities now in operation that possess such scrubbers, data indicates that the introduction of lime produces ash -- even the combined ash resulting from mixing bottom and fly ash -- which is highly alkaline; pH values of 11-12 or higher are typical (14,15).

Certain toxic metals -- most notably, lead -- are readily soluble in water under such highly alkaline conditions, due to their amphoteric nature: significant solubility at both low and high pH values. In tests of the ash from each of the U.S. facilities possessing scrubbers (14-17), and from similar Canadian (6) and Swedish (18) incinerators, lead and in some cases other metals have leached at high levels, often in excess of federal or state standards defining a hazardous waste, and often even when leached using distilled water or rain water, rather than the somewhat acidic medium employed in the EP test.

Recent Canadian studies have found a marked increase in the solubility of a wide range of organic chemicals present in fly ash as pH increases (19). These findings raise new concerns about the potential for acid gas scrubbers to enhance leaching of dioxins or other toxic substances from ash that are normally relatively insoluble in water -- a possibility that has yet to be tested at any U.S. facility possessing such scrubbers.

BIOAVAILABILITY AND DIRECT TOXICITY OF ASH

The hazards of ash must also be evaluated by direct bioavailability and toxicity testing, particularly with respect to the potential for ecosystem effects. Several studies demonstrate that toxic metals and dioxins present in ash are bioavailable to plants and animals (20-25), and direct ash toxicity -- attributable to toxic metals and/or dioxins -- has also been shown (2,5,26,27). In addition to the potential for direct environmental damage, these data document the plausibility of human exposure through contamination of the food chain.

Bioavailability is enhanced by the small particle size of a large fraction of ash, which allows direct inhalation or ingestion of such particles. Moreover, their small size promotes both short- and long-range dispersion, as is well documented for metal-containing particles released by various stationary and mobile sources (28-31), and has been demonstrated for MSW incinerators as well (32,33). These properties take on added significance in light of the permanent (metals) or persistent (dioxins) nature of the ash's toxins.

ROUTES OF EXPOSURE TO INCINERATOR ASH

Highly relevant exposure pathways exist not only from ultimate ash disposal, but from all earlier steps: onsite handling from the time of generation, storage, transport, and handling and depositing at the landfill until time of final cover. In each case, significant potential exists for both air-borne and water-borne dispersal of fugitive releases. Moreover, post-disposal exposure can occur as a result of direct ground or surface water contamination by leachate -- whether as a result of deliberate discharge of leachate, failure of the leachate collection system, a breach in containment systems, or the lack of maintenance of such systems that will inevitably follow the end of any required post-closure period. In addition, exposure can result from the handling and disposal of leachate or quench water (wastes which have themselves been found in some cases to contain toxic metals at levels near or exceeding hazardous waste limits -- refs. 4,14), as well as any residues generated through their treatment.

In addition to the permanent or persistent nature of ash-borne toxic substances, other properties of ash emphasize the need for long-term, secure containment. Studies have documented that under a range of circumstances, ash-borne metals can have significant mobility in soils, and that these metals can be taken up from ash-amended soil by plants (20,21,34). Thus, the potential for erosion of the final cap over time, transport of metals out of the landfill by plant uptake or other means, and anticipated end uses following landfill closure (e.g., as recreational areas) must all be seriously considered in assessing exposure routes.

Numerous studies on other wastes similar to ash such as metal smelter dusts demonstrate that improper management -- even long after initial disposal -- can result in actual human exposure to toxic metals through such pathways (29,35,36).

Perhaps the most graphic illustration of the lack of serious consideration of incinerator ash hazards is the fact that, to our knowledge, not a single quantitative risk assessment of incinerator ash has ever been conducted for a proposed incinerator project.

FUNDAMENTAL OBJECTIVES OF ASH MANAGEMENT

Given the clear hazards of ash, the major objective of any initiative to accomplish more environmentally sound ash management must be to reduce the hazardous character of the ash. In EDF's view, any approach to reducing the hazards of ash must provide for the development of strong incentives and regulations to accomplish the following: 1) separately test and manage fly and bottom ash; 2) dispose of ash separately from other wastes in secure facilities; 3) treat ash prior to disposal to reduce both its present and future hazards; and 4) keep toxic metals out of products that find their way into the municipal wastestream and keep materials containing such metals out of incinerators.

The first three objectives involve management at the back end, that is, after hazardous ash has been generated. While the last objective may appear to be beyond the scope of ash management, steps taken to remove metals from products or MSW prior to incineration are increasingly recognized as the economically and environmentally preferable methods of reducing ash toxicity. If metals were removed from trash prior to incineration, management of the resulting cleaner ash could be accomplished in a more protective manner than disposal of toxic ash even in a state-of-the-art landfill. Viable long-term solutions to the ash problem will require both stringent regulation of ash management and controls over the use of incineration that maintain opportunities and strong incentives to reduce ash toxicity at the source. Each of these objectives is discussed briefly below.

Separately Test and Manage Fly and Bottom Ash

There is clear and growing evidence indicating that for both economic and environmental reasons, separate management of fly and bottom ash is essential. Available data indicate that mixing actually increases the total amount of ash that can be expected to fail testing procedures and must be managed as hazardous waste. This is demonstrated by the fact that combined ash qualifies as hazardous waste more frequently than does bottom ash alone (see Table II).

Separate management of fly ash (which is only 5-15% of the total amount of ash) would reduce the total amount of ash that would have to be managed as a hazardous waste. In addition, mixing compromises our ability to effectively contain or treat toxic ash, since containment and treatment are more effective (and cost-efficient) when applied to a smaller volume/more concentrated waste.

Dispose of Ash Separately from Other Wastes

Co-disposal is recognized by virtually all parties, including the incinerator industry itself, to actually increase the hazards posed by disposal of incinerator ash. Recent data clearly indicate that the so-called buffering capacity provided by the alkaline nature of ash is insufficient to ensure long-term stability of toxic metals in disposed ash (6).

In EDF's view, monofilling -- the separate disposal of ash -- is a necessary but not sufficient component of proper ash management. While monofilling should be required, it does not supplant the need for full state-of-the-art containment with leachate collection and groundwater monitoring as additional essential design components. This need is even more apparent in light of the increasing use of acid gas scrubbers on new incinerators, which appears to enhance the leaching potential of lead even when water leaching is used to simulate the conditions of a monofill.

Encourage or Require Treatment of Ash Prior to Disposal

The need for treatment as well as separate management of fly ash prior to land disposal has been recognized in studies of ash contaminant leachability conducted by the Canadian government (6,37). A broad range of approaches (extraction, recovery, and fixation) is under investigation for treating incinerator ash to reduce metal leaching and the potential for fugitive releases of ash-borne toxins.

While such methods have considerable potential to reduce ash toxicity and should be aggressively pursued, it is equally critical that they be fully validated with respect to both their applicability to incinerator ash and their long-term as well as immediate effectiveness. Given the long-term hazards posed by metals in incinerator ash, treatment methods should be demonstrated to be effective under a range of conditions (e.g., multiple freeze-thaw cycles, pressure effects on structural integrity) that may occur even well beyond the end of the useful life of a disposal facility.

Ash treatment should be implemented through a treatment permit requirement, so that regulatory control over the means by which treatment is carried out can be exerted in order to ensure that the mandated treatment is carried out effectively and in a manner that provides full protection and containment during all management steps, up to and including ultimate disposal.

Keep Toxic Metals Out of Products and Keep Wastes Containing Metals Out of Incinerators

Strong incentives are needed to reduce the toxicity of ash at the source, through approaches such as product or process substitution, recycling, source separation, and preprocessing. The true costs of using and disposing of toxic materials in our trash are not reflected in a scheme that allows for less than fully protective management of ash residues. Such a scheme essentially codifies a subsidy for the continued use and improper disposal of such materials.

These same factors will also promote the continued use of the mass burn approach to incineration; this approach ignores the need to dissect the wastestream in order to use incineration only for those materials that can be safely burned. In EDF's view, the problem of toxic ash is a direct result of this blind approach to the use of incineration. In addition to providing oppportunites to remove materials that contribute toxic metals to the incinerated wastestream, preprocessing technologies can reduce the likelihood or frequency of upsets or reductions in combustion efficiency caused by trying to feed items to the incinerator that simply do not burn (38).

Even more serious, the growing reliance on mass burn incineration is acting to preclude use of recycling and other trash processing technologies that could serve to complement the use of incineration, to reduce the amount of incineration we must carry out, and to increase its safety with respect to both air emissions and ash toxicity. Recycling and source separation are essential elements in the safe and rational use of incineration. If we cannot move beyond the mere lip service paid to implementing and maximizing the use of these technologies that characterizes much of the incineration debate, we will simply perpetuate the same myth that brought us to the brink of the present landfill crisis: namely, that a single management technique can somehow manage our entire municipal wastestream.

LITERATURE CITED

1. Environmental Protection Agency, Background Document, Resource Conservation and Recovery Act -- Hazardous Waste Management, Section 3001, Identification and Listing of Hazardous Waste, Washington, D.C. (1980).

2. Knudson, J.C., "Study of Municipal Incineration Residue and Its Designation as a Dangerous Waste," Department of Ecology, Olympia, WA (1986).

3. Wilson, D.C., Young, P.J., Hudson, B.C., and Baldwin, G., Environ. Sci. Technol. 16, 560 (1982).

4. NUS Corporation, "Characterization of Municipal Waste Combustor Ashes and Leachates from Municipal Solid Waste Landfills, Monofills, and Codisposal Sites," 7 Volumes, EPA Contract No. 68-01-7310, prepared for U.S. Environmental Protection Agency (1987).

5. Kuehl, D.W., Cook, P.M., Batterman, A.R., Lothenbach, D.B., Butterworth, B.C., and Johnson, D.L., Chemosphere 14(5), 427 (1985).

6. Sawell, S.E., Bridle, T.R., and Constable, T.W., "Assessment of Ash Contaminant Leachability, NITEP Phase II (Quebec City)," Wastewater Technology Centre, Environment Canada, Burlington, Ontario (1986).

7. Norton, G.A., DeKalb, E.L., and Malaby, K.L., Environ. Sci. Technol. 20, 604 (1986).

8. Carlsson, K., Waste Management and Research 4, 15 (1986).

9. Greenberg, R.R., Zoller, W.H., and Gordon, G.E., Environ. Sci. Technol. 12, 566 (1978).

10. Wadge, A. and Hutton, M., Environ. Pollution 48, 85 (1987).

11. Brunner, P.H. and Monch, H., Waste Management and Research 4, 105 (1986).

12. Bridle, T.R., Cote, P.L., Constable, T.W., and Fraser, J.L., Water Sci. Technol. 19, 1029 (1987).

13. Brunner, P.H. and Baccini, P., "The Generation of Hazardous Waste by MSW-Incineration Calls for New Concepts

in Thermal Waste Treatment," Second International Conference on New Frontiers for Hazardous Waste Management, Pittsburgh, PA, Sept. 27-30, 1987.

14. State of Oregon, Department of Environmental Quality, "Extraction Procedure Toxicity Characterization of Municipal Incinerator Ash from Ogden Martin, Brooks," Claude H. Shinn, Laboratory Division, Portland (1987).

15. Resource Analysts, Inc., Test reports conducted for Signal Environmental Systems, Inc., dated 6 April, 30 May, 29 June, 1987.

16. Commerce Refuse-to-Energy Authority, letters to Dr. David Leu, California Department of Health Services, dated 11 May, 21 July, 12 November, 13 November, 1987, Whittier, CA.

17. Roy F. Weston, Inc., "Framingham Incinerator Ash Sampling and Analysis Report," for ESI, Inc., Southborough, MA, June 5, 1987.

18. Hartlen, J. and Elander, P., Swedish Geotechnical Insititute, "Residues from Waste Incineration: Chemical and Physical Properties," June 1986, Report No. 172, Linkoping, Sweden.

19. Karasek, F.W., Charbonneau, G.M., Reuel, G.J., and Tong, H.Y., Analytical Chemistry 59(7), 1027 (1987).

20. Wadge, A. and Hutton, M., Plant and Soil 96, 407 (1986).

21. Giordano, P.M., Bethel, A.D., Lawrence, J.E., Solleau, J.M., and Bradford, B.N. Environ. Sci. Technol. 17, 193 (1983).

22. van den Berg, M., Olie. K., and Hutzinger, O., Chemosphere 12(4-5), 537 (1983).

23. Van den Berg, M., Van Greevenbroek, M., Olie, K., and Hutzinger, O., Chemosphere 15(4), 509 (1986).

24. Van den Berg, M., de Vroom, E., Olie, K., and Hutzinger, O., Chemosphere 15(4), 519 (1986).

25. Opperhuisen, A., Wagenaar, W.J., van der Wielen, F.W.M., van den Berg, M., Olie, K., and Gobas, F.A.P.C., Chemosphere 15(9-12), 2049 (1986).

26. Kuehl, D.W., Cook, P.M., Batterman, A.R., and Butterworth, B.C., Chemosphere 16(4), 657 (1987).

27. Bronzetti, G., Bauer, C., Corsi, C., Del Carratore, R., Nieri, R., and Paolini, M. Chemosphere 12(4-5), 549 (1983).

28. Environmental Protection Agency, Air Quality Criteria for Lead, 4 vols., EPA-600/8- 83028aF-dF (1986).

29. Harper, M., Sullivan, K.R., and Quinn, M.J. Environ. Sci. Technol. 21, 481 (1987).

30. Nriagu, J.O., (ed.) Changing Metal Cycles and Human Health, Springer-Verlag, Berlin (1984).

31. Roberts, T.M., Hutchinson, T.C., Paciga, J., Chattopadhyay, A., Jervis, R.E., VanLoon, J., and Parkinson, D.K., Science 186, 1120 (1974).

32. Hutton, M., Wadge, A., and Milligan, P.J. Atmospheric Environment 21(10) (1987).

33. Berlincioni, M. and di Domenico, A., Environ. Sci. Technol. 21, 1063 (1987).

34. Mika, J.S. and Feder, W.A., "Resco Incinerator Residue Research Program: Results, Evaluations and Recommendations," University of Massachusetts, Waltham (1985).

35. Landrigan, P.J., Gehlbach, S.H., Rosenblum, B.F., Shoults, J.M., Candelaria, R.M., Barthel, W.F., Liddle, J.A., Smrek, A.L., Staehling, N.W., and Sanders, J.F., New England Journal of Medicine 292(3), 123 (1975).

36. Roels, H.A., Buchet, J., Lauwerys, R.R., Bruaux, P., Claeys-Thoreau, F., Lafontaine, A., and Verduyn, G., Environ. Research 22, 81 (1980).

37. Sawell, S.E., Bridle, T.R., and Constable, T.W., "Leachability of Organic and Inorganic Contaminants in Ashes from Lime-Based Air Pollution Control Devices on a Municipal Waste Incinerator," Air Pollution Control Association, June 1987.

38. Hershkowitz, A., Technol. Review, July 1987, 26-34.

39. Vogg, H., Braun, H., Metzger, M., and Schneider, J., Waste Management and Research 4, 65 (1986).

HOW CONTROL OF COMBUSTION, EMISSIONS AND ASH RESIDUES FROM MUNICIPAL SOLID WASTE CAN MINIMIZE ENVIRONMENTAL RISK

Floyd Hasselriis* ■ Gershman, Brickner and Bratton, Inc., Falls Church, VA 22043

Designers and operators of municipal waste combustors can achieve minimum environmental risk due to discharges of gaseous emissions to the atmosphere and solid wastes to recovery or landfilling, if they know what parameters need to be controlled, how to control them, and what monitors and measurements are needed to maintain control. This paper presents analyses of new data showing that both too much and too little excess oxygen or furnace temperatures cause increases in emissions of trace organics such as dioxins and furans; that oxygen is a critical measurement for control of combustion and carbon monoxide (CO) readings are important for monitoring of effective mixing of reactants; that acid gas scrubbers can condense organic and inorganic vapors which can be collected by fabric filters; and that either too high or too low pH can cause undesirable increases in the solubility of heavy metals such as lead and cadmium in ash residues. Careful operation and control can maintain optimum combustion, and minimize organic emissions and soluble heavy metals in the residues.

*Diplomate, American Academy of Environmental Engineers. Member: ASME, AIChE, APCA, ASTM; ASME Research Committee on Industrial and Municipal Waste.

Combustion is an efficient means of destroying the organic components of MSW and reducing it to a compact, inorganic ash residue. Modern technology provides means to destroy essentially all organics by combustion, remove most of the acid gases and almost all of the pollutants before the combustion products leave the stack, and collect them in the particulate. The bottom ash can be used for road-building and flyash has been used as an additive to concrete. An important issue today is how to minimize the soluble forms of heavy metals, especially lead and cadmium, which concentrate in the collected flyash.

The wide range in stack emissions and soluble metals in the ash from Waste-to-Energy (WTE) facilities has led to a public concern about why such variations exist, and, more important, how to achieve confidence that after the facility passes its compliance test, the emissions would also be acceptable on a hour-by-hour and day-by-day basis. This concern has led to the requirement for continuous monitoring of oxygen, carbon monoxide (CO), and acid gases in the stack gases, as well as furnace temperatures, and testing of the ash residues.

Extensive research on these emissions and discharges has been undertaken to determine how they can be reduced and maintained at environmentally acceptable levels, and, most important, to find which operating parameters could be measured and controlled in order to maintain these low discharges to the environment.

Environmentally safe incineration of MSW requires control of the combustion process, control of air emissions, and proper management of the flyash and bottom ash residues. In order for designers and operators to be able to accomplish this objective, they must have information on how to do it, and instruments capable of providing necessary measurements and control.

Municipal solid waste (MSW) contains organic and inorganic materials which will be affected in various ways by the combustion process and the emission controls. Whatever goes in must come out either with stack gases, flyash, or bottom ash. However, the chemical form may be significantly different.

Organic materials containing sulfur, nitrogen and chlorine are converted to carbon dioxide, water, acid gases, and trace organics. With good combustion, toxic organics in stack emissions, collected flyash, and bottom ash can be reduced to extremely low levels.[1] Vendors of MSW combustion systems have developed sophisticated equipment to accomplish these objectives.[2] However, there is a continuing need

*Gershman, Brickner and Bratton, Inc., Falls Church, Virginia, 22043.

to define the conditions required to achieve this.

The challenge to chemists and chemical engineers is to discover and understand what factors influence the chemical form of the toxic heavy metals so that the process of combustion, emission control and ash residue treatment can assure an environmentally safe product. The major part of the ash residues from combustion of MSW can be used beneficially, and need not relegated to landfills if processed into safe materials.

Tests of MSW burning plants show a wide range of emissions of polychlorinated dibenzo-p-dioxins (PCDD) and polychlorinated dibenzofurans (PCDF), as shown in Table 1. These trace toxic organics may enter with the waste, be created in poor combustion or form in post-combustion zones under certain conditions. Diagnostic tests have shown that there are significant relationships between combustion conditions and the emission of PCDD/DF before and after emission controls, and that mixing effectiveness, tightness of control, moisture, furnace and post-furnace temperatures, and the use of lime and reduced temperatures for acid gas control all have an effect on emissions of trace organics.

Parameters such as temperature and/or oxygen, and the use of CO as an indicator, have been suggested as means to find and maintain optimum combustion conditions so as to minimize dioxins and furans.[6] It has been found that acid gas controls can reduce emissions below those achieved by good combustion alone.[7]

As the ability to measure lower levels of emissions developed, and emissions were reduced to even lower levels, it has become more important to have instrumentation which could reliably measure and control these low levels.

The development and use of reliable oxygen measurement and control devices for use in improving fuel economy in fossil fuel-fired boilers has advanced the use of these relatively rugged devices in WTE systems. The main difficulties have been protecting them from the high moisture and dust content of gaseous emissions from burning MSW. These problems have been reduced by heating piping connections, providing filters and means to purge the filters.

Fossil fuel control systems have long used carbon monoxide monitors to discover and maintain the optimum oxygen level which would correspond with minimum CO emissions. The question is, does low CO correspond to low emissions of trace organics such as dioxins? If so, then CO monitors and oxygen control can be used to assure minimum emissions of these toxics.

DIOXIN AND FURAN EMISSIONS

CAUSES OF DIOXIN FORMATION

Many explanations for the formation and destruction of dioxins have been put forth. They can enter in the MSW, be created in cold regions of a furnace, be destroyed in combustion, form in the cooler outlet sections of the boiler, and 'all of the above.'

Vogg and Stieglitz have found from extensive laboratory research as well as field tests on a MSW incinerator that:

o   Formation of PCDD and PCDF takes place at temperatures ranging from 200°C to 400°C. Figure 1 shows no effect on dioxins and furans at temperatures below 200°C, a sharp peak in both dioxins and furans which occurs at 300°C (570°F), and their destruction at 400°C (800°F).

o   In this temperature range formation leveled off after about six hours, but in one half hour about 20% conversion had taken place.

o   Oxygen concentration influenced formation of PCDD/DF linearly: zero oxygen resulted in decomposition or no formation; increasing oxygen levels resulted in a reduction in the fraction of dioxins and furans having the more toxic four-chlorine forms (congeners) and an increase in fraction of less toxic forms having 6 to 8 chlorine molecules attached.

o   Moisture strongly influenced dechlorination, causing formation of the highly toxic penta and tetra (4 and 5-chlorine) isomers.

o   Only trace PCDD/DF were detected in boiler flyash deposits in the second and third passes of the boiler as the gases cooled from 1470 to 750'F, but substantial amounts were found in pass four where the gases cooled from 750 to 425°F.

o   Copper chloride ($CuCl_2$), together with the alkali/alkaline chlorides in the fly ash, appears to play an important catalytic role, releasing free chlorine in reactions which take place on carbon surfaces.[9]

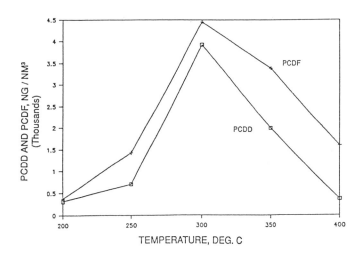

Fig. 1 Variation of dioxins (PCDD) and furans (PCDF) formed in flyash in range from 200° to 400°C. Vogg et. al. [9]

o The hydrochloric acid (HCl) is oxidized from alkali and alkaline earth chlorides (such as KCl), also releasing chlorine to react with carbon.

o The amount of carbon in the flyash appears to affect formation of PCDD/DF directly by the well-known Deacon process by which HCl is oxidized to Cl_2 with airborne oxygen. Carbon can be reduced by good combustion, and is probably a useful indicator for dioxins and furans.

Vogg concludes that good combustion, reduction of precursors such as carbon, as well as keeping the boiler surfaces clean, are the primary measures which can be used to minimize dioxins, and suggests that ammonia could be used to poison the catalysts for the PCDD/DF reaction.[9]

The broad range in dioxins and furans reported from tests of various plants is shown in Table 1, including CO and furnace temperature readings when available. An enormous reduction in these emissions has been reported at recent plants: from 10,000 down to less than 1 nanogram per cubic meter.

TESTING AND RESEARCH ON DIOXINS

Tests at many Waste-to-Energy (WTE) plants have gradually brought to light the relationships between operating parameters and emissions of carbon monoxide (CO) to organic emissions, including dioxins and furans. Since these toxic compounds can only be measured by lengthy and expensive testing, it is essential that we find out what measurements of operating conditions or surrogates could be used to control and supervise the combustion system to assure low levels of dioxins and furans an a continuous basis.

Analysis of data from diagnostic testing of the Hamilton, Ontario (SWARU) plant revealed that the quantity of dioxins and furans emitted from combustion of refuse-derived fuel (RDF) were related to operating parameters such as furnace temperature, excess oxygen, air distribution and emissions of CO and dioxins.[10]

Various tests of the Hampton, Virginia plant showed that while higher temperatures and lower CO were associated with reduced dioxins, data from one set of data showed a reverse tendency: higher temperatures produced higher dioxins under conditions of plant overload, with highly oscillating furnace conditions. Other test data reported by Benfenati also showed this tendency of increased dioxins at higher temperatures.[6]

The tests of the Prince Edward Island plant, under the Canadian NITEP program, revealed that although dioxins and furans were reduced to low levels, only a weak reduction in dioxins was noted as temperatures were varied from the lowest to the highest practical temperatures for refractory, starved-air incinerators.[11]

The ASME Joint Dioxin Committee decided that detailed research was needed in order to answer many questions about the combustion of MSW. The ASME/NYSERDA Pittsfield test program was designed to explore the relationships between waste composition (MSW, commercial trash (called PVC-free), PVC and moisture), and combustion parameters; especially to determine the relationships between trace organic emissions and CO, temperature, oxygen and moisture on emissions of dioxins and furans. It was planned to explore the entire range of temperatures within which the incinerator could operate in order to find the optimum operating conditions, and how they could be maintained.[12]

This facility had features which made it especially suitable for full-scale research:

o refractory furnaces produce more constant temperature gradients than water-cooled furnaces, hence temperature becomes an easy-to-measure parameter;

TABLE 1. PCDD, PCDF AND CARBON MONOXIDE EMITTED BY VARIOUS PLANTS

Plant Location	Type of Plant	Type of Control	Condition	Stack Emissions, ng/nM^3			CO	TEMP.
				PCDD	PCDF	Total	ppm	deg.C
Hampton, VA,	MSW	ESP	max	13,000	24,000	37,000	1000	
		ESP	min	670	3,700	3,670		
Hamilton, ON,	RDF	ESP	max	1,700	7,000	8,700	480	738
		ESP	min	1,300	4,000	5,300	300	764
Chicago, IL	MSW	ESP	max	61	490	551	140	
Albany, NY	RDF	ESP	avg	300	88	388	200	
PEI, ONT	MSW	None	max[1]	123	156	279	40	788
			min	64	100	164	14	1038
Neustadt, WG	MSW	ESP/SCR	in[2]	80	95	175	—	
			out	5	9	14	—	
Stapelfeld, WG	MSW	ESP/SCR	max	40	120	160	—	
			min	20	90	110	—	
Peekskill, NY	MSW	ESP	avg	18	40	58	30	
Pittsfield, MA	MSW	ESP	Ph I	76	270	346	140	677
(Research tests)			Ph II	55	144	200	144	700
			"	8	17	25	30	760
			"	0.7	5	5.7	9	843
			"	0.8	10	10.8	7	843
			"	14	36	50	5	980
			"	28	39	67	14	980
			Ph I	24	38	62	16	1010
Wurzburg, WG	MSW	SCR/FF	max	36	54	90	40	
			min	12	10	22	31	
Tulsa, OK	MSW	ESP	avg	19	19	36	16	
Bristol, CT	MSW	SCR/FF	avg	2.4	6.2	8.6	18	
Marion, OR	MSW	SCR/FF	max	1.5	2.0	3.5	17	
			min	0.8	1.0	1.8	14	
Biddeford, ME	RDF	SCR/FF	avg	1.2	1.6	2.8	83	
Montreal, Q[4]	MSW	ESP	March	0.75	0.54	1.2	nm	
			Nov.	0.01	0.02	0.03	nm	
Quebec, Ont.	MSW	SCR/FF	264°C[4]	1,300	1000	2,300	220	
			200°C	6.1	1.2	7.3	160	
			140°C	0	1	1.0	220	
			125°C	nd	nd	nd	180	
			110°C	0.2	2.3	2.5	140	

1. Lowest and highest practical furnace temperatures.
2. In and out of wet scrubber (TUV, 1984)[3]
3. Lime fed with the MSW.(Boisjoly)[4]
4. In and out of dry lime scrubber.(Hay, 1986)[5]

and boiler inlet temperatures; hence both oxygen and temperature can be independent parameters within some range;

o Combustion air and recirculated gases are effectively mixed with combustion gases, minimizing the variable of mixing.

Preliminary (Phase I) tests were run to test out the instrumentation and analytical methods, including sampling for dioxins and furans at the minimum and maximum practical furnace temperatures of about 1500 and 2000°F. These tests showed the close relationship between the readings of CO and temperature at temperatures below about 1700°F, and gave indications of a rise in CO at temperatures above 1700°F. On the basis of these findings, a series of temperatures was planned for the test matrix of the Phase II tests.[12]

DIOXIN TEST RESULTS

The Pittsfield test data confirm the relationships between temperature, oxygen and CO with dioxins and furan emissions which had previously been noted, and added a great deal of detail to this information. The report stated that adding PVC to the refuse produced no statistically significant change in dioxin emissions, and that conditions were optimum when the furnace was operated within a range of temperature and oxygen.

The effect of secondary furnace temperature, measured after complete mixing of secondary air, on carbon monoxide (CO) is shown in Fig. 2, and on total dioxins (PCDD) plus furans (PCDF) in Fig. 3. Logarithmic plots have been used to spread the data and to show the logarithmic relationship between these variables.

At temperatures below 1700°F CO is noted to have a closely linear relationship with temperature, as would be expected from chemical kinetics. However, at higher temperatures a substantial rise in CO is noted. By comparison, PCDD+PCDF showed a more erratic yet similar trend at temperatures below about 1600°F, and a more substantial increase at temperatures above 1750°F. The two runs (24 and 29) with moisture added to the PVC-free waste show substantially lower PCDD+PCDF levels, and levels roughly equal to those found for the high-moisture MSW tests (runs 15 and 21). The same comparison can be noted in Fig. 2 for CO.

It should be noted that the temperatures reported were secondary furnace exit temperatures, after admission of secondary air which included flue gas recirculation. The actual flame temperature, estimated by calculating the cooling effect of the secondary air, was about 150 to 200°F higher. Hence the optimum 1600°F furnace temperature corresponded with a flame temperature of about 1800°F, and the temperature above which emissions increased was about 2000°F.

Effect of oxygen, moisture and PVC. The effect of oxygen on PCDD+PCDF is shown in Fig. 4, where PCDD+PCDF exhibit a minimum at about 9% oxygen, increasing on either direction, but sharply as oxygen is increased. Excess oxygen (or air) varies inversely with temperature for refuse having the same moisture content. Runs with high moisture had consistently low dioxins and furans, whereas runs with added PVC showed increased furans when oxygen was reduced below about 9%.

The linear plot in Fig. 5 shows that PCDD+PCDF decreases as CO falls to the range of 10 to 20 ppmv, but increase sharply below that. The log-log plot of the same data, shown in Fig. 6 shows the rapid increase in PCDD+PCDF below CO of 30 ppmv, and the scatter of data from runs with different kinds of waste. The wide vertical variation shows that other variables are present. Note that the high moisture (H_2O) runs had the lowest PCDD+PCDF, and that the both PVC-free and PVC-added data points are scattered in the group at low CO with high PCDD+PCDF. There is no statistical difference between runs with different waste composition: the trend is strongly related to the CO level, however.

Comparison between Pittsfield and Amager data. A comparison with data from the Amager plant in Denmark shows trends similar to the Pittsfield data in spite of the difference in technology. Tests of this excess air refractory incinerator with a Volund rotary kiln, consisted of 12 runs: #2,3,4 during startup, #5,6,7 under normal conditions, #8,9,10 with sludge burning, and #11,12,13 at abnormally high loads.[13]

Figure 7 shows a linear plot of the Amager data showing that both CO and PCDD+PCDF follow a logarithmic approach toward zero CO as excess oxygen is reduced. No data was obtained at conditions below 11% oxygen, which would probably have been more favorable. Organic emissions were high during startup (runs 2,3 and 4), lower during normal and sludge-burning tests, and best at high load, except for the outlier, run #11.

The results of these tests show that CO and dioxin emissions are closely related, that the nature of the refuse and the operating

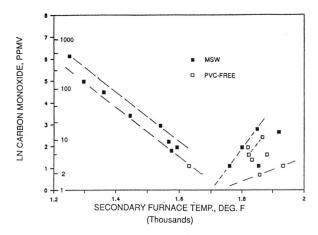

Fig. 2 Logarithm of carbon monoxide versus furnace exit temperature, showing all test data. MSW points filled in, open points are commercial waste. MSW, MSW plus PVC, PVC-free (paper and cardboard), with and without PVC addition, and high moisture were run in duplicate. Data from Visalli.[12]

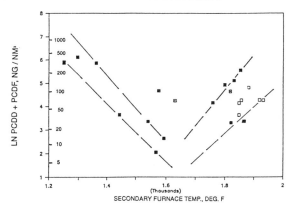

Fig. 3 Pittsfield data: Logarithm of total dioxins (PCDD) plus furans (PCDF) versus temperature measured at exit of secondary furnace. A linear relationship is noted at temperatures below 1650°F.

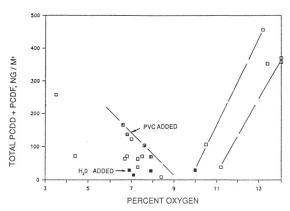

Fig. 4 Total dioxins plus furans (PCDD+PCDF, ng/M^3) versus excess oxygen measured at outlet of secondary furnace, showing rise at oxygen levels above or below 9.5%.

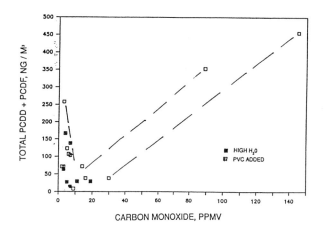

Fig. 5 Total PCDD+PCDF plotted versus carbon monoxide (CO). Curves drive toward zero but increase occurs below about 10 ppmv.

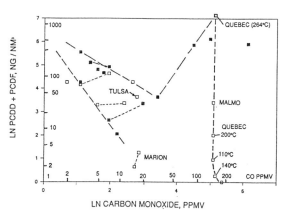

Fig. 6 Logarithm of Total PCDD+PCDF versus carbon monoxide (CO). While curves drive toward zero, a sudden increase occurs below about 10 ppmv. Quebec acid gas control data are shown.

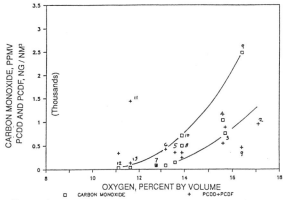

Fig. 7 Amager data: Carbon Monoxide and PCDD+PCDF vs. excess oxygen. Tests 2,3 and 4 were startup; 5,6 and 7 normal operation, 8,9 and 10 during sludge burning, and 11,12 and 13 were high load operation. [13]

conditions of the incinerator are important to the extent that they affect temperature, available oxygen, and proper air distribution.

Combustion Efficiency versus oxygen. The relationship between excess oxygen and Combustion Efficiency, calculated from CO, is shown in Fig. 8. Optimum CE values around 99.995% were attained at 7 to 8% oxygen.

Process oscillations. Some data points fall out of the generally low range, probably due to oscillations in operating conditions from too much air to too little air. The average of such a range of conditions includes a high percentage of poor operating conditions, resulting in high emissions. Figure 9 shows the envelopes of scatter-gram data recorded during Pittsfield test runs at different set temperatures. Even though the temperatures were held very closely, CO varied substantially. To hold 1250°F, more air (oxygen) had to be added as needed to compensate for moisture variations. Likewise, to hold 1850°F, it was necessary to reduce the oxygen to about 3%, causing high CO levels. The most stable, low-CO runs took place at 1550°F, measured in the outlet of the secondary furnace. The corresponding flame temperature in the primary furnace, prior to mixing with combustion air and recirculated gases, calculated from gas flow measurements, is about 1700 to 1750°F. It appears that oxygen is more effective for control than temperature.

Figure 10 shows a trend-graph from continuous monitoring of a controlled-oxygen incinerator. Normally the oxygen was held within the range from 6 to 8%. An upset is shown, which was out of normal range for less than 20 minutes. Oxygen ranged up to 13% and down to 5.2%, with total recovery about an hour later.

In the EPA Report to Congress an explanation was given of why tests of the Hampton facility in Virginia revealed such high dioxin levels: the combustion air damper was oscillating from wide open to closed, rather than seeking a more stable condition.[2] The published data shows oxygen ranging from 3 to 12% during most of the runs, in the same manner as shown in Fig. 9.

CONTROL OF ACID GASES, ORGANICS AND INORGANICS

Control of temperature

Temperature is a major factor in the effectiveness of emission controls. At temperatures below about 280°F sulfur dioxide and mercury, as well as dioxins, furans and heavy metals, are effectively removed, primarily by condensation on particulate.

Dry powder injection and spray-dry scrubbers have both been shown to be effective in removing acid gases and metals at these temperatures. This has been confirmed by exhaustive tests of a pilot plant attached to the Quebec City incinerator.[5] These tests, performed under the National Incinerator Testing and Evaluation Program (NITEP) show extremely low emissions of dioxins measured after the scrubber/fabric filter when the gas temperatures were held below 140°C (280°F). Even though the incoming PCDD and PCDF were high, in the range of 1,000 ng/m^3, indicating relatively poor combustion, the outlet was below detection limit or at most 1 ng/m^3. The removal efficiency was thus higher than 99.9% for dioxins and furans at 140°C, when using either dry lime injection or spray-dry (lime slurry) scrubbers.

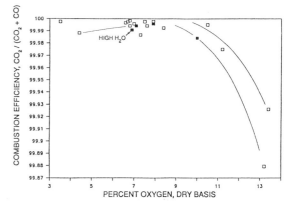

Fig. 8 Pittsfield data: Combustion efficiency versus excess oxygen. Substantial decreases are noted on either side of optimum oxygen. Note tests with high moisture.

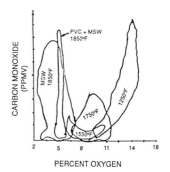

Fig. 9 Carbon monoxide (CO) measured by continuous monitor versus excess oxygen at boiler outlet. Secondary furnace temperatures were held at noted temperatures during each test run, shown by envelope of scattergram points. One test included added PVC.[12]

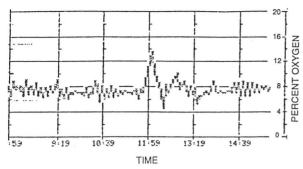

Fig. 10 Controlled oxygen measured at furnace outlet of furnace similar to Pittsfield. An upset in the feeding cycle shown, deviating from the normal pattern of control. Scale is 0-20% oxygen.

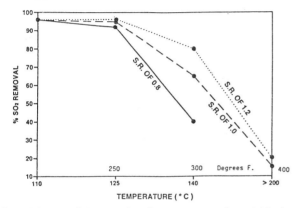

Fig. 11 Sulfur dioxide removal efficiency versus scrubber temperature at different excess lime (stoichiometric) ratios (S.R.), from NITEP tests of spray-dry and dry lime injection scrubbers. Below 125°C the removal efficiency exceeds 90% regardless of S.R.

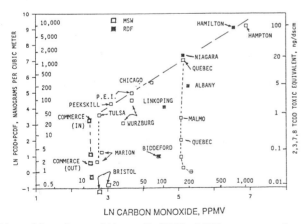

Fig. 12 Logarithm of PCDD+PCDF versus logarithm of carbon monoxide. Pittsfield data before emission controls is compared with Tulsa data after an ESP, data from Quebec City and Commerce before and after lime scrubber and baghouse, and Marion and Biddeford data after scrubber/baghouse.

Table 2 shows the relationship between temperature and the collection efficiency of heavy metals and mercury. Table 3 shows the removal efficiencies of acid gases. It is apparent that reductions in temperature of the combustion products drastically increase control effectiveness.

Effect of temperature and lime stoichiometry

The NITEP tests show that while increasing the quantity of lime used in scrubbing has an effect on removal efficiencies at temperatures over 300°F, at lower temperatures, especially below 250°F, even as low a stoiciometric ratio of 0.8 was almost as effective in achieving over 90% SO_2 control as an S.R. of 1.0 or 1.2, as shown in Figure 11. This finding is extremely important, not only in achieving high removal efficiencies, but also in economising in lime. Even more important may be the avoidance of excessive lime in the ash residues, since excess alkalinity can act to mobilize heavy metals.

It should not be overlooked that as the stack gas temperature is reduced, the plume rise of the gases leaving the stack will be reduced, perhaps substantially increasing ground level concentrations. For this reason, and also for reasons of practical difficulties with equipment due to condensation of acids and water vapor, there is a practical optimum stack temperature, below which disbenefits become significant in regard to benefits. Stack gas reheat may be required to improve plume rise and reduce ground level concentrations.

Dioxin control by combustion vs. condensation

Emissions of dioxins and furans can be reduced by two methods: good combustion, and effective condensation and collection by emission controls. Recently tested facilities having acid gas scrubbers, such as Marion County, have demonstrated extremely low emissions of dioxins, while operating at low CO levels.[14] The tests of the City of Commerce, CA, Bristol, CT, and Biddeford, ME, facility show remarkably similar emissions, in spite of relatively high CO emissions at Biddeford.[15]

In the tests of the un-retrofitted Quebec City facility, the high levels of dioxin emissions were reduced to extremely low levels by applying a lime scrubber, indicating that if the belt fails, the suspender still works.

After retrofitting the Quebec City incinerator, with modifications to the furnace and combustion controls, extremely low dioxin and furan emissions were achieved without resorting to scrubbing and temperature reduction of stack gases.[16]

The separate effect of good combustion and effective scrubbing and particulate control with reduced gas temperatures is shown in Fig. 12. The general trend of PCDD+PCDF with CO for plants with ESP's reflects the improvement in combustion as CO is reduced to about 10 ppm.

The effect of the emission controls is shown in the vertical line containing the data from the Quebec City tests. In this case, with CO in the range from 160 to 220 ppm, the gases leaving the boiler had 1500 ng/Nm^3 of PCDD+PCDF, whereas leaving the baghouse the emissions ranged down to about 1 nanogram or not detectable (as shown in Table 1), dramatically demonstrating the effectiveness of the scrubber and baghouse.

The tests of the Biddeford refuse-derived fuel (RDF) facility show similar low dioxin and furan emissions, but at much higher CO levels (80 ppmv, average), indicating that even with relatively high CO the scrubber is the ultimate PCDD/DF removal device.

When the combustion is optimum, as is the case with the Tulsa and Marion plants, CO was about 15-20 ppmv, but with the ESP only, PCDD+PCDF were about 40, as compared with about 2 ng/Nm^3 from the Marion facility which has a scrubber/baghouse. The difference between these two plants may be attributed to the scrubber/baghouse. The total amount of dioxins and furans collected by the flyash is much less in the case of the plants with effective combustion.

SOLUBLE METALS IN ASH RESIDUES

The heavy metals present in MSW are collected in flyash or bottom ash except for trace quantities which are emitted with the stack gases. Most of these metals are in a chemical form which is not soluble in natural water or rain. However, there is concern about the small fraction of these metals which, if the ash were not managed properly, could be dissolved in water or leachate which might enter the environment at unacceptably high concentrations which could contaminate surface water or ground water, possibly drinking water.

Published data on leaching properties of ash residues have shown a wide variation. Although most bottom ash and mixed ash residue samples have passed regulatory tests, flyash has generally failed due to high lead and cadmium solubility. There is little data from operating plants which have acid gas scrubbers, which change the properties of the flyash and mixed ash due to absorption of acids and the presence of excess lime.

In order to be assured that ash residues are handled so as to represent no significant threat to the environment, it is necessary to understand how they can be properly managed, so that they may be safely disposed of, or used beneficially, thus reducing the quantities which must be landfilled.[8]

THE EP TOXICITY TEST

The Extraction Procedure (EP) Toxicity test was developed in order to simulate a worst case method of ash disposal: in an acid-producing MSW landfill on the assumption that about 5% ash residue would be mixed with 95% MSW or other wastes which, it was assumed, could generate a leachate with a pH of 5 as they decompose. If the ash sample failed this test, it would be considered risky to dispose of this ash in such a landfill, on the assumption that leachate from the landfill would leak into groundwater. The EP test, and the Toxic Concentration Leaching Procedure (TCLP) test developed for the Resource Conservation and Recovery Act (RCRA), use an acetic acid solution to 'force' the leaching by maintain a pH close to 5 to rapidly extract the metals.

The amount of heavy metal extracted during the EP and TCLP tests depends upon the buffering ability of alkaline components of the ash residues which resist the acid added during the test. The American Society for Testing Materials (ASTM) procedure and methods used in Europe use simulated acid rain or distilled water in an effort to simulate conditions where the residue is used in road construction or placed in cells separated from MSW in a landfill. Such tests would be more appropriate for measuring the potential effect on the environment of these methods of use or disposal of ash residues.

CONTROLLING LEACHABILITY OF HEAVY METALS IN ASH

When the RESCO facility in Saugus, MA started operation (1975), it was planned to find uses for the screened ash residues rather than landfilling them. A large section of highway was constructed using these residues. However, plans to use them for landfill cover and other uses were frustrated by concern about the occasional samples of ash residues which did not pass the EP toxicity test.

TABLE 2. REMOVAL EFFICIENCIES OF HEAVY METALS AT 140 C[1]

METAL	AVERAGE INLET (ug/Nm^3)	OUTLET RANGE (ug/Nm^3)	REMOVAL EFFICIENCY (percent)
Zinc	100,000	6	99.995
Lead	40,000	1 - 5	99.99
Cadmium	1,125	ND - 0.4	99.97
Chromium	2,200	0.2 - 1	99.97
Nickel	1,600	0.7 - 1.3	99.95
Antimony	1,500	0.3 - 0.6	99.97
Mercury	400	10 - 20	97.50
Arsenic	115	0.04	99.97

1. Source: D.J. Hay et al. Range reflects performance of dry powder and spray-dry scrubber performance at 140°C.[5]

TABLE 3. REMOVAL EFFICIENCIES OF ACID GASES [5]

POLLUTANT		INLET (ppmv)	OUTLET (PPMV)	% REMOVAL EFFICIENCY
SO_2	(140°C)	100	41	58
	(125°C)	118	10	92
HCl	(140°C)	475	29	94
	(125°C)	464	9	98

A comprehensive testing program was initiated by RESCO-Saugus in order to resolve questions about the high variability of ash residues from this plant. The testing program, much of which was carried out at the University of Massachusetts, included studies of plant uptake of heavy metals and leaching of ash residues in simulations of land application. When the wide variations were noted, test programs were set up to discover the causes of variation, as well as the potentials for treatment of the residues. In addition, the ash residue storage pile was investigated by boring over 30 feet into the pile. Much of this has been reported by Feder and Mica.[17] Recently this ash pile has again been tested, and the ash has been found to pass the EP tests and produce extremely low metals concentrations in the leachate.

Some major findings of this research:

o   Heavy metals condense on boiler tubes as the gases cool; metal-rich deposits fall into the bottom ash during tube cleaning, causing spikes in EP test samples.

o   Heavy metals condense on fine particulate, collected mainly in back sections of the electrostatic precipitator (ESP).

o   Rapping the boiler and ESP caused surges of dust to be collected which, when discharged into the bottom ash, caused occasional imbalances in the residue composition, and spikes in the data.

o   Leaching tests of flyash showed that the acidity or alkalinity (pH) of the leaching water strongly influenced the solubility of lead and cadmium. At excessively high acidity or alkalinity, these metals became highly soluble. The minimum solubility of lead was found to be in the pH range from about 8 to 10, with a sharp increase with either higher or lower pH leachate.

o   Because of the sensitivity to the amount of acid added during the EP toxicity test, it was found that the results depended on who performed the test.

o   Two or three washes were found to remove almost all of the lead and cadmium from the ash, indicating that most of the soluble metals were on the surfaces.

o   The mixed flyash and bottom ash stored in a pile on the site was found to have lost its leachable metals down to about 16 feet below the surface, leaving a 'clean' ash residue.

Lysimeter (leaching column) tests showed that if the leaching water stays within the pH range of low metals solubility, leaching will be slow and the leachate may have a low and non-hazardous metals concentration. Since the ash residues are strongly alkaline, leachate will maintain the alkalinity which results in slow leaching even when large amounts of weakly acid 'acid rain' are passed through it. By the time that the ash has begun to lose its buffering ability, hardly any of the leachable metals would be left. This has been demonstrated by long-term simulations of ash leaching using ash residues from the Westchester, NY, WTE facility.[18]

VARIABILITY OF LEACHING PROPERTIES OF RESIDUES

The degree of mixing, or the effectiveness of a mixing process can be measured by the standard deviation of the data. Figure 14 shows plots of EP toxicity measurements of samples of bottom ash, mixed ash and flyash. The mixed bottom ash shows the characteristic of both the flyash and the bottom ash. The flyash which is normally fairly well mixed shows a steeper slope, similar to RDF ash, and lower standard deviation than the mixed ash. This graph is a log/Rosin-Rammler graph, slightly different from a log-probability graph.

Points which lie along a straight line represent approximately equal probability of occurring in a purely random situation. The graph is constructed by placing the data in increasing order, and plotting the points at equally divided intervals of 100%. If the data follow the line closely, they are indeed random with equal probability. Points which fall off of the line may be analytical outliers, data from different populations, or relatively unlikely points. If the outlier is forced onto the straight line by moving it up, it will reveal its probability, which, for instance, may be 1 in 100, whereas the other data may be 1 in 10 (for 9 points). This technique is useful in visualizing and evaluating the effectiveness of any process which involves variables, including mixing.[19]

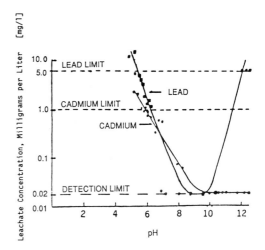

Fig. 13 Solubility of lead and cadmium of acid gas scrubber plus flyash product, as a function of the pH of the leaching solution, by TCLP method. Note that a pH of 5 causes the lead and cadmium to exceed the EPA limits, whereas at a pH between 8 and 10 solubility is below the detection limit of 0.02 mg/l. Above 10 lead solubility increases and exceeds the limit at pH = 12.[20]

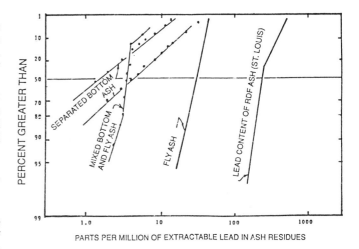

Fig. 14 Variability of Extraction Procedure toxicity test data for flyash and bottom ash samples collected during single days, plotted on Rosin-Rammler paper, and compared with EP soluble lead content of flyash and total lead content of shredded municipal solid waste (RDF).[18]

LEACHING OF METALS FROM SCRUBBER PRODUCTS

In order to predict the composition and leaching properties of residue from an acid gas cleaning system planned for WTE projects, extensive pilot testing has been performed.

scrubber were tested in order to predict the ash residue characteristics of the RDF-fired WTE plant in Hartford, CT. The results of these tests showed that combining scrubber products with flyash produced a material which passed the EP toxicity test for most metals including cadmium and lead. Apparently the scrubbing process resulted in rendering these metals immobile.[20]

To develop methods for use and disposal of scrubber and flyash products, a pre-testing program was carried out during the design phase of the Southeast Massachusetts WTE facility. Flyash was collected from an operating refuse-derived fuel (RDF)-fired boiler, and various simulated products were made by blending this flyash, a spray-dry absorber (SDA) product from an operating system in Zurich, Switzerland, and calcium hydroxide together in several blends. A low-ash SDA product was collected from the pilot plant at Niro's Copenhagen research facilities. Stabilized products using aluminous cement, Portland cement, granulated blast furnace slag and cement kiln dust were blended in various mixtures and tested. After curing for two weeks, cast prisms were tested for porosity and compression strength. Crushed stabilized products were analysed for the eight RCRA heavy metals after leaching according to the proposed TCLP procedure. Figure 13 shows how the lead and cadmium concentrations of samples of SDA acid-gas control system product mixed with flyash varied with the pH of the mixture.[21]

When lime is used for acid gas control, it is equally important that excessive alkalinity not be developed by poor process control, since pH levels over 11 cause lead to become soluble.

These tests show clearly how strongly the leaching properties of flyash are affected by the use of lime in acid scrubbers. Appropriate management of lime scrubbing residues, flyash and ash residue handling systems makes it possible to control the fraction of heavy metals which is soluble, so that these residues will not be classified as hazardous wastes. If properly managed, these residues may be safely disposed of, or used beneficially, thus reducing the quantities which must be landfilled.[22]

A recent study of the ash residues from the Marion County WTE facility which employs a spray-dry lime scrubber, concludes that the "EPtox and TCLP extractions are nor a very realistic representation of the leaching processes the ash is exposed to after disposal in a monofill... A more realistic extraction procedure would be multiple extractions using real or synthetic rainwater. The use of acetic acid or acetate buffers is meaningless considering the disposal in "monofill," except that pH reduction is hastened compared to the use of rainwater as an extraction fluid."[23]

Recent data on leachate from the Marion County ash-fill which receives mixed lime scrubber flyash and bottom ash shows that the pH of the actual leachate is in the neutral range from 7 to 9.

CONCLUSIONS

The composition of MSW does not appear to affect the emissions of dioxins and furans from combustion: rather, temperature and effective mixing of reactants are the more important parameters.

Furnace temperature is an important parameter; however, temperature increases beyond some limit, which may be specific to the technology, show increases in organic emissions. The Pittsfield data show that at secondary furnace temperatures (after fully mixing secondary air) greater than 1600°F, and specifically from 1750 to 1950°F, PCDD+PCDF increased substantially above the minimum point. Prior to adding secondary air the primary furnace temperatures were 150 to 200°F higher than the secondary furnace temperature. The increase in emissions at higher temperatures may be the result of the reduction in excess oxygen which is required to achieve higher temperatures.

Carbon monoxide readings are an indicator of good combustion within a certain range. However, while CO can be reduced to extremely low levels by increasing temperature, dioxin and furan emissions have been noted to rise significantly at CO levels below about 10 ppm. Other limiting parameters such as excess oxygen, or mixing intensity, must also be used to define optimum operating conditions.

Oxygen measured after combustion is a primary control parameter, directly related to furnace temperature and fuel moisture. High oxygen and excess air reduce the flame temperatures; too little oxygen makes it more difficult to mix the reactants, and uniformity of destruction of organics suffers.

During the Pittsfield tests, combustion efficiencies ($[CO_2]/[CO_2 + CO]$) exceeding 99.995% were achieved at minimum dioxin levels when oxygen was held at 7 to 8%, dry basis.

Stable control of combustion is important in order to prevent oscillations into conditions of poor combustion. Measuring and controlling excess oxygen within a narrow range appears to be the best way of maintaining optimum combustion conditions.

Results of tests of the Amager rotary kiln incinerator in Denmark are consistent with the Pittsfield data, indicating that the relationships between operating parameters and emissions of dioxins and furans are uniform and combustion-related.

Post-combustion control of acids, heavy metals and organics by reducing temperatures, scrubbing with lime and collecting particulate on a fabric filter, are effective in controlling emissions even when combustion is not ideal; however, good combustion reduces the quantity of organics which are collected on the particulate.

Consistent control of lime addition to match acid concentrations can prevent production of excess acidity or alkalinity in the scrubber and flyash products.

Failure to mix flyash with bottom ash in reasonably constant ratios can result in a non-uniform product which may not consistently pass the Extraction Procedure (EP) toxicity test.

Less lime is needed to achieve SO_2 control at scrubber temperatures below 125°C (250°F).

The amount of lead and cadmium which can be leached from flyash and bottom ash depends upon the design of the combustion system and boiler, the operating conditions, the method of mixing flyash with bottom ash, the amount of excess lime, and water chemistry.

Leaching tests which apply deionized water or simulated acid rain appear to simulate landfill leaching conditions better than the EP and TCLP tests which apply acid which counteracts the natural buffering ability of the ash residues.

The natural alkalinity of the ash residues, including excess lime from acid gas control, buffers the weak acidity of acid rain and inhibits the release of lead and cadmium.

Good process control to maintain optimum conditions can produce quality-controlled ash residue which can be used to assure products which would not be classified as hazardous, thus permitting their recycling and use rather than landfilling.

SUMMARY

Concern about emissions and ash residues from combustion of MSW has led to research on methods to improve combustion controls, reduction in emissions, and process control and treatment of ash residues. Parameters and surrogates which can be used to find and maintain minimum emissions and residue toxicity have been found as the result of diagnostic testing, and can now be used to monitor, supervise and control the operation of waste-to-energy facilities.

o effective destruction of organics by combustion can be assured by use of excess oxygen and furnace outlet temperature measurement and control, combined with carbon monoxide measurements which are a surrogate of destruction efficiency of organics;

o low scrubbing temperatures are a surrogate for the effective condensation of acid gases and heavy metals removed from combustion products by fabric filters which effectively remove essentially all particulate, condensed metals and acids;

o leaching of heavy metals from the naturally alkaline ash residues from the incineration of MSW, can be kept to minimum levels by assuring that they will not be subjected to excessing acid or alkaline environments;

o alkalinity control, effective mixing of ash components, and control of chemical reactions and treatment can be used to minimize solubility of heavy metals in flyash, scrubber and bottom ash residues;

o process control, based on a knowledge of optimum conditions, as well as effective mixing of reactants, is the key to finding and maintaining minimum environmental discharges;

o proper design, management and operation of facilities to process and burn MSW makes it possible to reduce environmental discharges to extremely low levels, reducing risk below the risks associated with landfill and other methods for disposal of wastes.

REFERENCES

1. Hay, D.J., A. Finkelstein and R. Klicius, "The National Incinerator Testing and Evaluation Program: An assessment of A) Two-stage Incineration and B) Pilot-scale Emission Control," APCA Annual Meeting, Minneapolis, MN, June 1986.

2. Seeker, W.R., W.S. Lanier and M.P. Heap, "Municipal Waste Combustion Study: Combustion Control of MSW Combustors to Minimize Emission of Trace Organics," EPA/530-SW-021c, June, 1987.

3. TUV, "Tests of Uncontrolled and Clean Gas and Discharged Salts of a MSW Incinerator for PCDD/PCDF," Hamburg, FRG, 1984.

4. Boisjoly, L., "Measurement of Emissions of PCDD and PCDF from Des Carrieres Incinerator in Montreal," Report EPS 5/UP/RQ/-1, Environment Canada, Dec. 1984.

5. NITEP, "Air Pollution Control Technology," Report EPS 3/UP/2, Sept. 1986, Environment Canada, Ottawa, Ontario.

6. Hasselriis, F., "Minimizing Trace Organic Emissions from Combustion of Municipal Wastes by the Use of Carbon Monoxide Monitors," 1986 ASME SWPD Conference, Denver, June, 1986.

7. Hasselriis, F., "Minimizing refuse combustion emissions by combustion control, alkaline reagents, condensation and particulate Removal," Synergy/Power Symposium, Washington, October, 1986.

8. Hasselriis, F., "How Ash Residues Can Be Used Safely," ATSWMO Conference, Los Angeles, 1987.

9. Vogg, H., M.Metzger and L.Steiglitz, "Recent Findings on the Formation and Decomposition of PCDD/PCDF in Solid Waste Incineration," Waste Management & Research (1987), 5.

10. Hasselriis, F., "Optimization of Combustion Conditions to Minimize Dioxin Emissions," WHO/ISWA Seminar, Copenhagen, Jan. 1987; Waste Man. & Res. (1987), 5.

11. NITEP, "Two-stage Combustion (Prince Edward Island)," EPS 3/UP/1, 1985, Environment Canada, Ottawa, Ontario.

12. Visalli, J., "Results of the Combustion and Emissions Research Project at the Vicon Incinerator Facilility in Pittsfield, MA," NYSERDA, Albany, N.Y., Report 87-16, June 1987.

13. Grove, Arne, "Emissions of Dioxins From Incineration of MSW at Different Operating Conditions," Waste Man. & Res. (1987), 5.

14. Sussman, D., and J.Hahn, "Dioxin Emissions Test Results from Two New Municipal Waste Combustors," Waste Man. & Res. (1987), 5.

15. McDannel, et. al., "Air Emissions Tests at Commerce Refuse-to-energy Facility, May 26-June 5, 1987," San. Districts, Los Angeles County, Whittier, CA, July 1987.

16. NITEP, "Mass Burning Tech., Quebec City-Phase II, Envir. Canada, June, 1988.

17. Feder, William A. and Jane S. Mika, "Summary Update of Research Projects with Incinerator Bottom Ash Residue," Executive Office of Environmental Affairs, Commonwealth of Mass., Feb. 1982.

18. Cundari, K.L., "The Laboratory Evaluation of Expected Leachate Quality from a Resource Recovery Ashfill," Dept. of Public Works, Westchester Cty. NY, 1986.

19. Hasselriis, F., "Variability of Composition of Municipal Solid Waste and Emissions from its Combustion," ASME Solid Waste Proc. Conf. Miami, 1982.

20. Borio, D.C. and A. Plumley, "Tests of Scrubber Baghouse Pilot Plant," Kreisinger Res. Lab. Combustion Engineering, Windsor, CT, 1986.

21. Donnelly, James R. and Ebbe Jons, "Byproduct Disposal from MSW Flue Gas Cleaning Systems," APCA, New York, 1987.

22. Shinn, Claude E., "Toxicity Characteristic Leaching Procedure (TCLP), Extraction Procedure Toxicity (EPtox) and Deionized Water Leaching Characteristics of Lead from Municipal Waste Incin. Ash," DEQ Lab. and Appl. Res., Portland, OR, July 1987.

23. "Toxicity Characteristic Leaching Procedure (TCLP), Federal Register, Vol 51, No. 216; Nov. 7, 1986, Pp. 40643-40654.

24. Environmental Test Report, Ogden Projects, Inc., Report Number 130, January, 1988.

25. Barnett, Vic, and Lewis, "Outliers in Statistical Data," John Wiley, 1978.

TRENDS IN FINANCING PUBLIC WORKS: EVALUATING PRIVATIZATION OF MSW DISPOSAL FACILITIES

Mark R. Herron, Kline P. Barney, Jr., ■ Parsons Municipal Services, Inc., Pasadena, CA 91124

The financing of public works is a dynamic process, and each segment competes with other public needs for available funds. Recent changes in the tax treatment on invested money requires an astute evaluation and careful planning. The success of past projects suggests that government can accommodate changes to encourage infrastructure reconstruction. Case studies illustrate the variety of funding techniques available and changes in tax treatment are proposed to encourage private investment.

INTRODUCTION

This paper will explore the options available to public agencies (cities, counties, and regional authorities) for the financing of public works infrastructure projects. In recent years, government at all levels has found it harder and harder to raise money to fund important projects. There are a number of reasons for this, including the perception on the part of many voters that government is inefficient and the only way to control spending is to limit government's access to funds in the first place. (California's "Proposition 13" is just one well known example.) This is occurring at the same time much of our existing infrastructure is wearing out and increased concern over the environment are causing us to questions certain past practices.

One of the financing options available to public agencies is the use of "public/private partnerships"--partnerships between the public and private sectors formed to meet a particular public sector need. This has also been termed "privatization." Privatization takes many forms, from the simple "contracting-out" of services (such as trash collection), to the private development of large capital facilities and even to the sale of government assets (such as the Conrail railroad) to the private sector.

The handling of municipal solid waste (MSW) is one of the greatest challenges facing public agencies today. The shear volumes of MSW that are being generated every day require that a solution be found. But old practices (such as landfilling) are being questioned. Alternate technology is available, but it is more expensive--often very much more expensive--than old solutions. This paper will look at how both conventional and alternate financing methods can be applied to MSW disposal projects.

FINANCING OPTIONS

Until recently, public works infrastructure projects were almost always financed and owned by public agencies such as cities, counties or regional agencies responsible for such vital services as wastewater treatment, MSW disposal or water supply. The option of private ownership and financing is a fairly recent development.

The availability of tax-exempt debt to public agencies make public financing (and thus public ownership) particularly attractive. This form of financing is the most common and probably will continue to be. (Tax-exempt debt is debt issued under certain provisions of the federal Internal Revenue Code. The interest payments to those who buy the bonds are excluded from income subject to income tax at the federal, and in certain cases, the state level. Because the interest income is exempt from taxation, the debt can carry a lower interest rate and still be attractive to lenders. The public agency issuing the debt benefits from a lower cost of funds.)

The Tax Reform Act of 1986 placed certain restriction on the ability of local government agencies to issue tax-exempt debt. The most important is that no more than ten percent of the proceeds of a bond issue be used "directly or indirectly by a trade or business." Most public projects qualify under this requirement, although a large industrial user or commercial establishment using a facility could bring into question the facility's status.

Public agency debt can be either in the form of "general obligation" debt or a "revenue" obligation. General obligation debt is just what the name implies: the issuing agency is obligated to make interest and principal payments from any and all sources of funds available to it. Revenue bonds, on the other hand, rely on the revenue produced from a specific source. This source is usually directly related to the facility being financed. For example, the revenue from the sale of potable water is often pledged to pay for water treatment facilities. Because a general obligation pledge is more secure than a pledge of a specific revenue source, revenue bonds usually carry a higher interest rate to compensate bond holders for the additional risk.

A variation of traditional municipal debt is a municipal lease. The interest component of the lease payment is exempt from federal (and possibly state) tax, just as it is on municipal debt. Municipal lease financing is more commonly used for financing such assets as vehicles and computers, although it has been used to finance more expensive assets with longer useful lives. (Parsons Municipal Services recently arranged lease financing with a term of twenty years for a $33.5 million cogeneration facility for the County of Los Angeles.)

The funding of lease payments on most municipal leases is usually subject to annual appropriations by the public agency leasing the assets financed. This distinguishes a lease from debt. Although there are a number of protections that make the appropriation of funds almost certain, there is an element of risk making this sort of financing more expensive. On the positive side, a lease is usually not considered debt and thus lease obligations are excluded from most debt limitations imposed on local jurisdictions by state law. They also do not usually require approval of voters as most general obligation debt does.

Another source of funds available to local public agencies has in the past been grants, usually from the federal government but sometime from the state. Most federal grant programs are being phased out or greatly reduced. Many communities are finding the grant process almost more trouble than it is worth. The local share of project cost has been increased, grants take years to obtain, and the conditions ("strings") applied to the grant monies are complicated and expensive to administer.

A final source of funds that may become more important in future years is state revolving loan programs. Several states have set up such programs, which can lower the cost of borrowing to local agencies.

Private financing for infrastructure facilities comes in two basic forms: debt and equity. The debt issued to finance a project is usually revenue debt. The lenders look to certain project revenues to pay interest and repay principal. The revenue can be a fee from users (a bridge toll, for example) or from a service fee paid by public agency (such as a service fee paid on a wastewater treatment plant). In certain circumstances, the debt can be tax exempt: "exempt facility bonds" (formerly called industrial development bonds) can be issued on behalf of a private company for certain types of facilities.

Equity or "risk capital" will be invested in a project where there is a reasonable expectation that the funds left over after debt service and other expenses are paid will flow to equity investors and provide a market rate of return on their investment. In the past a large part of an equity investors return came from the tax benefits (accelerated depreciation and investment tax credit) available on the assets financed. These tax benefits are to a large extent no longer available, as explained below. Because tax benefits are not available to public entities (because they do not pay tax in the first place) this form of return on investment was essentially without cost to the public entity. In the post tax reform environment, return on any equity investment must come from other sources.

CHANGES RESULTING FROM THE TAX REFORM ACT OF 1986

As mentioned above, the Tax Reform Act of 1986 made changes in the Internal Revenue Code which made public debt somewhat more complicated and expensive to issue. The major impact of the new law was to make private financing less attractive. The primary reason for this is the substantial reduction or elimination of tax benefits available on the assets of large capital projects. Depreciation (the permissible write-off of the value of an asset for tax purposes) has been stretched out. The reduced tax rates make depreciation write-offs less valuable. And investment tax credits, formerly as much as ten percent of the value of the asset, have been eliminated altogether.

The tax benefits provided a large part of the return to equity investors necessary to attract investors to public infrastructure projects. Because these benefits were available, the overall cost to the public agency was reduced.

The other significant change resulting from the Tax Act of 1986 is a significant restriction on the availability of tax-exempt financing through industrial development bonds, now called exempt facility bonds. The projects that qualify for such financing have been limited and a state-by-state cap on the dollar volume of exempt facility bonds issued has been made more restrictive. However, facilities for the processing of solid waste are excluded from the cap limits, although they must meet the other requirements of the new law. In addition, a strict (and probably unrealistically low) limit has been placed on the portion of bond funds that can be expended on costs of issuance (legal and investment banking fees, for example).

WHAT IS THE FUTURE FOR PRIVATIZATION OF MSW FACILITIES?

Privatization is a cooperative effort. It must truly be a public/private partnership. Privatization is not and has never been a way to finance otherwise uncreditworthy projects. In the past, privatization did enable a public agency to receive what was in effect an indirect grant from the federal government in that tax benefits not otherwise available to the public agency were made available. The private sector is willing to assume certain project risks if it can expect to receive a reasonable return on the investment it makes to assume these risks. This is particularly true in the case of MSW facilities.

Providing facilities for the disposal of MSW is a traditional role of government. It is not often looked at from the perspective of risk-return analysis, and thus it is difficult for many public officials to understand the concerns of the private sector when considering public/private partnerships as a possible solution. Many times the public sector will attempt to place all the risk of a MSW facility on the shoulders of the private sector when trying to structure a privatization arrangement and expect the private sector to provide services at cost comparable with existing alternatives. This is just not possible.

However, if the public sector takes the cooperative approach and works with the private sector to reach a solution, viable MSW disposal options are possible. The technology exists, the private sector is willing to assume reasonable risks (with the expectation of a reasonable return, and the private sector can bring efficiencies and innovation that may not otherwise be available to the public sector.

MANAGEMENT OF LEACHATE FROM SANITARY LANDFILLS

Jeff M. Harris and James A. Gaspar ■ Browning-Ferris Industries, P.O. Box 3151. Houston, TX 77253

Sanitary landfills remain the most common form of solid waste disposal in America. However, environmentalists are concerned about the degradation of ground water by leachate, which is generated in the landfill by precipitation percolating through and mixing with the waste fill. The resulting leachate is usually a high strength wastewater. This paper provides an overview of modern landfill design technology to minimize leachate formation, prevent ground water quality degradation, and dispose of the leachate in an environmentally responsible manner.

INTRODUCTION

Solid waste is an unavoidable by-product of human existence and activities. Each American produces an average of 5.5 pounds of solid waste per day. Proper disposal of this waste stream is imperative. The most common method of disposal is by sanitary landfilling. One by-product of landfilling is the generation of a liquid waste known as leachate. This paper will define leachate and methods of collecting, treating and disposing of the waste liquid.

Collection and treatment of sanitary landfill leachate has a limited past history. The field of knowledge is in a continual state of evolution and improvement. Thus, today's "state of the art" system may soon be obsolete. Similarly, state and Federal regulations vary greatly in regard to leachate. Some states, such as New Jersey, have very specific design requirements, other states simply require that leachate be collected, while still others have no definitive requirement. Despite this variability, it is obvious that the trend is for virtually all new landfills to have leachate collection capability. Additionally, many closed landfills will require remediation to deal with leachate. This dictates the necessity of leachate treatment and disposal alternatives.

The management of leachate from sanitary landfills will be a significant concern for years to come.

LEACHATE DEFINITION/COMPOSITION

Definition of Leachate

Leachate is a complex organic liquid formed primarily by the percolation of precipitation water through the open landfill or through the cap of the completed fill. To a lesser extent, leachate can be formed as a result of the initial moisture content of the waste. The resulting leachate is a complex and highly variable mixture of soluble organic, inorganic, and bacteriological constituents and suspended solids in an aqueous medium. The exact composition of leachate is variable and site specific, depending on the type of refuse accepted at the site, the age of the refuse, and the amount of precipitation. Compared to municipal wastewater with COD's of 250-1000 mg/L and TSS's of 100-400 mg/L, sanitary landfill leachate is a high strength wastewater with COD's up to 90000 and TSS up to 1000 mg/L. Leachate compositions are highly variable; some typical BFI leachate characteristics are listed as well in Table 1.

Leachate composition variance is sometimes explained by the type of wastes in the fill. For example, the leachate may contain pathogens because of buried disposable diapers and/or "red bag" hospital wastes. Sanitary landfills constructed prior to enactment of the RCRA guidelines for hazardous waste control may contain industrial and/or hazardous waste. If large quantities of plaster or gypsum (calcium sulfate) were buried in the fill, the anerobic activity would convert much of the sulfate to sulfide. The leachate would appear black because of iron sulfide and could cause sulfide odor problems. Depending on the local industries using the landfill, may contain high levels of toxic organics and/or heavy metals. The most important factor affecting leachate composition is landfill age. According to Chian and DeWalle, leachate from young landfills contains high levels of BOD due to the presence of volatile fatty acids (acetic, propionic, and butyric). These compounds are the products of primary anaerobic decomposition of waste, specifically poteins, fat, and carbohydrates are shown in Figure 1. They also tend to be easily degraded so that young leachates (about 2-5 years old) are characterized by high BOD/COD ratio, higher pH, a more biodegradable leachate, and lower heavy metals concentrations (because the metals are less soluble at the higher pH).

Total dissolved solids (TDS) concentrations decrease with age as well, according to Costa. Apparently, the initial slug of acidic leachate dissolves the majority of soluble salts; as the landfill ages and the leachate pH increases, the net salt content is depleted and calcium salts become less soluble.

Because of the variability in leachate composition, the leachate should be well characterized prior to treatment process selection. If a sample cannot be obtained (for example, if the landfill is new and leachate is not being produced yet), data from nearby facilities may be used; however, varification of these results, when possible, is prudent. Table 2 lists the recommended characterization parameters and their significance to facility design.

FACTORS AFFECTING LEACHATE PRODUCTION

The best leachate management technique is to minimize its formation by proper landfill design. However, leachate generation in landfills is complex and depends on several independent variables. Figure 2 summarizes many of the factors affecting leachate generation 4, grouped according to:

* availability of water in the landfill;
* landfill surface conditions;
* refuse conditions; and,
* underlying conditions.

Designers and operators of landfills address these factors to minimize leachate generation and its environmental risks. If not managed responsibly, the risk of groundwater pollution increases and may lead to an expensive, long-term remediation effort to restore the groundwater's original purity.

LANDFILL DESIGN

A proper understanding of leachate management techniques requires a basic knowledge of sanitary landfill as a waste disposal method. Current landfill design encompasses civil, environmental, geotechnical and hydrogeological engineering principles to minimize leachate production and its environmental risks. The design is highly scrutinized by experts in the field, as well as by regulating authorities during the permitting phase. Public hearings provide the opportunity for citizen review and comment. It currently requires two years or more from initial concept to groundbreaking for a typical sanitary landfill site. Figure 2 summarizes the major components of the design and permitting of a landfill site. Figure 3 illustrates a typical cross section, and Figure 4 shows a schematic drawing of an operating sanitary landfill.

The design of a modern landfill generally begins with a site selection process followed by an extensive field investigation. The field investigation includes drilling and soil sampling to determine the suitability of the site for land disposal. Important parameters include:

* soil stratigraphy beneath the site;
* laboratory characterization of the soil permeability;
* safe excavation depth; and,
* other parameters to determine the suitability of the proposed site for waste disposal.

Concurrently, a hydrogeological study is done to develop background information on groundwater. This may include the installation of piezometers to determine the groundwater elevation and monitoring wells to allow collection of groundwater samples. Wells are often placed in a 15-60 meter buffer zone between the toe of the proposed landfill and the property line. Samples are periodically collected and analyzed over a period of up to one year to provide a groundwater quality baseline which will be used to determine any future degradation caused by the landfill.

Landfill construction begins with an excavation to a predetermined safe working depth based on seasonal high groundwater levels. The side slopes of the excavation are typically laid back to a 3 horizontal to 1 vertical slope to prevent slope failure. Depending on the type of soils, a steeper or shallower slope angle may be specified. The design criteria is generally based on maintaining a minimum safety factor of 1.5 against slope failure. The bottom and sides of the landfill excavation are then covered with a recompacted liner of natural clays. At the proper moisture level, clay may be compacted to a dense and impermeable state which is ideal to contain leachate. The clay is generally compacted so that the in situ permeability is not greater than 1×10^{-7} centimeters/second. If suitable natural clays are not available on site, the material can be brought in from off-site borrow areas, or in some cases the existing soil can be amended by addition of bentonite. It is also possible to use a synthetic membrane such as PVC or HDPE, or some combination of natural soils and synthetics to achieve the desired permeability. Some states now require a combination of clay liner and synthetic membrane known as a composite liner. Quality Control/Quality Assurance is critical for placement of the liner system since this forms the primary shield against environmental degradation by the outflow of leachate or the inflow of groundwater.

Upon completion of the liner and leachate collection system, garbage is spread in lifts, compacted to a minimum density of 1000 lbs/c.y., and covered with at least 6 inches of compacted soil at the end of each operating day. The soil cover minimizes odor and rainfall absorption. Current practice is to continue the fill well above ground level. This maximizes the volume of a landfill site and is economically necessary due to the extreme cost and lengthy time frame of permitting a new site. Above-ground side slopes are most commoly placed at 1 vertical to 4 horizontal (25%) with a typical top cap slope of 2-5%.

The completed portions of the landfill are capped with compacted clay soil and sometimes synthetic membranes to minimize rainfall infiltration. The uppermost layer is generally topsoil to support vegetative growth. Vegetation serves several purposes including improved aesthetics, erosion control, and increased evapotranspiration (evaporation and transpiration of water from vegetation) which helps reduce water levels within the completed landfill. The design of the landfill includes surface water management features such as diversion channels and let down strucures to route water away from the fill and prevent this water from infiltrating the fill. A network of groundwater and landfill gas sampling wells is also specified. Some landfills include perimeter slurry walls to cut off the flow of groundwater into the site. A slurry wall is a vertical trench cut deep enough to key in to an underlying impermeable layer and backfilled with a bentonite-stabilized soil slurry. The bentonite swells and creates an impermeable wall to prevent groundwater intrusion and/or leachate migration.

LEACHATE PRODUCTION DETERMINATION

One common task in the design and operation of a landfill is to predict the quantity of leachate that can be generated. Estimation of leachate production is important to:

* optimize the excavation and filling sequencing of a landfill by keeping the open excavated area to a minimum;
* size the leachate collection system;
* size on-site leachate treatment facilities or estimate the required capacity of off-site treatment plants;
* predict the potential for environmental impact of a release to the environment;
* determine the most effective final cap (clay cap vs. clay plus synthetic composite); and,
* calculate the financial reserves necessary for post closure leachate treatment.

Several mathematical methods have been used to estimate the volume of leachate generated by landfills.[9] All are based on the water balance principle, a methodology originally developed for the soil and water conservation fields. Figure 5 illustrates the components of a landfill water balance. The water balance method is a one-dimensional flow model based on the law of conservation of mass which states that the amount of mass input into a system must equal the mass output from the system plus the change of mass stored within the system. The major elements of a moisture balance are:

* precipitation, P;
* potential evapotranspiration, PET;
* surface runoff, R/O;
* infiltration, I;
* soil moisture storage, ST;
* change to soil moisture storage, ST;
* infiltration minus potential evapotranspiration, I-PET;
* Neg (I-PET), the sum of the negative values of I-PET;
* actual evapotranspiration, AET;
* percolation, PERC.

The percolation is then estimated by the formula:

Perc = P - R/O - ST - AET.

The percolation as calculated is analogous to leachate for a landfill situation. This model is based on several assumptions including:

* the sole source of infiltration/percolation is precipitation falling directly on the landfill's surface. Runoff from adjacent areas cannot be added to the balance.
* no groundwater enters the landfill.
* water movement through the landfill is vertical.
* The landfill material is at field capacity.
* leachate recycling is not allowed.

The time for first appearance of leachate can also be estimated. Assuming that channeling of the flows does not occur, leachate will appear when the landfill cap and the refuse mass reach field capacity. Field capacity is the maximum moisture content which a soil or refuse mass can retain in a gravitational field without producing continuous downward percolation.

A further development of the leachate generation prediction models came about with the EPA HELP, or Hydrologic Evaluation of Landfill Performance computer model. The following description of the HELP model is taken from the User's Guide: "The Hydrologic Evaluation of Landfill Performance (HELP) computer program is a quasi-two-dimensional hydrologic model of water movement across, into, through, and out of landfills. The model accents climatologic, soil, and design data and utilizes a solution technique that accounts for the effects of surface storage, runoff, infiltration, percolation, evapotranspiration, soil moisture storage, and lateral drainage. Landfill systems including various combinations of vegetation, cover soils, waste cells, special drainage layers, and relatively impermeable barrier soils, as well as synthetic membrane covers and liners, may be modeled. The program was developed to facilitate rapid estimation of the amounts of runoff, drainage, and leachate that may be expected to result from the operation of a wide variety of landfill designs. The model, applicable to open, partially closed, and fully closed sites, is a tool for both designers and permit writers."

The model is based on modeling the layers of a landfill from the cap down through the liner system. Default climate data is built into the model for about 100 cities. Specific data can be entered for cities not in the default database. Soil data is also built into the model for a variety of soil types. Again, specific data for a particular site can be manually entered if none of the default values are appropriate. The output provides the user with average monthly, average annual, and peak daily estimates of the various components of precipitation, including the leachate generation rate. The HELP model is available through the National Computer Center and is available on disks for running on a PC.

TREATMENT OBJECTIVES

The goal of leachate treatment is to reduce the concentration of pollutants to allow discharge into surface waters (streams, rivers, lakes), or to pre-treat constituent concentrations to acceptable levels for transfer to an off-site treatment plant.

For discharge into surface waters, the effluent concentrations will be dictated by NPDES permit conditions set by local regulatory agencies. The permit conditions depend on stream classification and the ratio of proposed effluent flow to low stream flow conditions.

For eventual discharge to an off-site treatment facility, maximum effluent concentrations are generally limited by pretreatment ordinances dictated by the treatment works. These ordinances are designed to prevent plant upsets, capacity reduction, sludge quality degradation, and odor problems. Most POTW's surcharge customers based on BOD, TSS, and/or nitrogen levels above a base value; therefore, there is economic incentive to reduce the organic and nitrogen loadings via pretreatment.

LEACHATE TREATMENT TECHNOLOGY

A process sketch of a hypothetical leachate treatment system is shown in Figure 7. The individual unit processes include:

* flow equalization;
* precipitation;
* VOC and/or ammonia stripping;
* neutralization;
* anaerobic biological treatment;
* aerobic biological treatment;
* sedimentation/filtration;
* chemical oxidation; and
* carbon adsorption.

Each operation is briefly discussed.

Flow Equalization. Equalization ponds are used as surge tanks to prevent shocks or upsets to downstream operations, especially biological processes whose performance is easily affected by changes in organic levels. They also allow uniform flow loading of downstream treatment units. Ponds are frequently aerated with minimal air requirements to avoid odor, maintain a homogeneous mixture and accomplish some biologic stabilization. Studies have indicated a 20-40% reduction in BOD in aerated equalization ponds.

Precipitation. High heavy metals concentrations can inhibit biological activity or exceed downstream requirements (set by NPDES permit or off-site treatment plant). Metals which are usually targeted for control include As, Cu, Zn, Cd, Pb, Ni, Ag, Hg, Cr, and Fe.

Precipitation via lime addition appears to be a good process choice as long as chelating agents are not present. Precipitation via sulfide addition (using FeS or NaHS) is an alternative to hydroxide precipitation. Advantages include:

* reduced chemical costs because of lack of need to raise pH;
* improved cake dewatering characteristics; and
* achievement of lower residual metals concentrations, particularly for Cu, Ni, Zn, Cd, and Pb.

Unit operations of a precipitation system include:

* reagent slurry tank and feed pump;
* reagent/leachate mixing zone,
* sedimentation tank with optional coagulant/flocculant addition to aid settling; and
* filtration to dewater the cake.

The entire process can be done continuously or treated as a batch process within the same tank. Sludges from the process are potentially hazardous and should be disposed according to regulations.

Precipitation of heavy metals is a complex chemical process; chelant and electrolyte interaction effects can cause wide variation from theoretical equilibrium predictions. Screening tests are recommended to quantify reduction, sedimentation, and cake handling criteria.

Stripping. Biological growths can be inhibited by high concentrations of ammonia (greater than about 500 ppm) or excessive concentrations of volatile organic compounds (VOC's). VOC's are a class of insoluble organic and organo-halide chemicals. Both ammonia and VOC's can be removed via air stripping. Air-liquid contact can be provided via shallow retention ponds or standard packed tower scrubbers.

If the leachate is acidic, ammonia stripping must be preceded by pH adjustment to convert ammonium cation (non-volatile) to free ammonia (volatile).

Neutralization. If pretreatment processes included metals precipitation with base addition, it will probably be necessary to re-acidify the leachate to an optimum pH prior to biological treatment (typically 6.5 to 7.5). Though leachate is usually nearly neutral, it may be desirable to include neutralization capability to control potential excursions.

Equipment requirements include acid storage, chemical feed pump, and mixing. On-line pH monitoring or control can probably be used with confidence because of its relatively low cost and high reliability. Phosphoric acid can be used simultaneously to neutralize and to provide nutrients for downstream biological treatment.

Anaerobic Treatment. Anaerobic biologic processes utilize microbes cultivated in the absence of oxygen to degrade BOD to sludge, carbon dioxide, and methane. Advantages over aerobic treatment include lower sludge production, lower cost of operation because aeration is not required, and the ability to handle higher levels of specific toxic organics. However, anaerobic processes are sensitive to thermal (low temperature) changes and BOD load shocks, accomplish no nitrification (conversion of ammonia nitrogen to nitrate nitrogen), and are difficult to start-up (requires bacterial "seeding" with manure or sludge).

The most important unit operation for industrial anaerobic treatment is the anaerobic filter, in which wastewater flows through a fixed media coated with microbes. The media provides a rigid and high-area surface for the acid-forming and methanogenic microbes to permanently attach to (as opposed to suspending the biomass by turbulence). Woelfel, et. al., described a successful installation of an upflow anaerobic filter for leachate treatment in Wisconsin. The filter proved stable operation for 6 months at BOD loadings less than 100 lb/day/1000 ft^3. The authors reported 90% BOD and TSS reductions. Cu, Pb, and Zn removals were 80%. The metals apparently accumulated and were removed in the filter's produced sludge.

Aerobic Biological Treatment. Aerobic biological processes utilize microbes cultivated in an oxygen rich environment to oxidize BOD to CO_2, water, and a suspension of cellular matter. Figure 8 shows a typical activated sludge aerobic system. At some point, the cellular matter, or activated sludge, is allowed to settle. A portion of the sludge is purged from the system and landfilled, while the remainder of sludge is recycled to maintain high concentrations of cell tissue. Sludge from municipal landfill leachate treatment plants may contain high levels of toxic heavy metals and require disposal as a hazardous waste. Frequently, phosphate and/or nitrogen nutrient addition is required to support the biological growth. Start-up usually requires an acclimation period to develop a successful microbe population, and the initial BOD loadings should be low enough to allow population growth.

Aerobic unit operations include:

* aeration basins;
* activated sludge recycle; and
* sequencing batch reactors (SBR).

Aeration basins are aerated lagoons with large detention times; they are inefficient and used where large tracts of land are available. They are generally not suitable for cold climates due to the inhibitory effect of cold weather on biological activity. A potential enhancement to aerobic basin treatment is the use of lemna plants to consume BOD, TSS, sulfides, nitrogen, and other pollutants. Lemna Corporation has developed a potential system to control the plants' growth and to harvest the mature plants.

In activated sludge and SBR's, less retention time is required because activated cellular tissue is recycled to speed the degradation rates. The SBR is useful because it utilizes the same tank to aerate, settle, and recycle solids. The SBR is a batch processing device; the sequence of batch steps to effect a cycle is shown in Figure 9. It is an optimum choice for low flow applications such as leachate treatment because it reduces the number of required equipment items.

Sedimentation/Filtration: Sedimentation and filtration processes are used to remove suspended solids, stabilized sludge, or precipitate from leachate.

Leachate may contain high levels of TSS (1000 to 50000 mg/L) compared to domestic wastewater at 100-350 mg/L. Removal of the TSS prior to biological treatment will reduce the BOD loading on a biological process. Sedimentation is a simple process in which sufficient time is given to allow solids to settle; addition of flocculants (alum or polyelectrolytes) will cause small particles to coagulate and settle faster.

Filtration is normally used as a pretreatment to prevent plugging of carbon adsorption beds or to further reduce TSS levels. Back-washable sand filters are usually employed.

Chemical Oxidation. Chemical oxidants such as chlorine, chlorine dioxide, sodium hypochlorite, potassium permanganate, and ozone may be used to reduce BOD and COD. Other applications include:

* selective destruction or alteration of toxic organics;
* disinfection; and
* enhanced denitrification.

Ammonia nitrogen is toxic to fish and so may be regulated in discharge permits. Its removal via biological treatment occurs in two steps, nitrification (conversion of ammonia to nitrate - nitrogen in aerobic units) and denitrification (conversion of nitrate-nitrogen to nitrogen gas in special anoxic units). However, ammonia may be directly converted to nitrogen gas via breakpoint chlorination. The economics should be considered when ammonia removal is a problem.

Chemical oxidants are expensive to use; they will be cost effective only with hard to treat leachates with low BOD/COD ratios and low levels of organics.

Activated Carbon. Activated carbon is produced by thermally activating coal particles to a high porosity. Soluble organic molecules selectively adsorb onto the high surface area created by the activation. Nonpolar molecules tend to adsorb more readily; thus activated carbon is ideal for removal of many VOC and other toxic chlorinated compounds.

Spent activated carbon is regenerated by hot air stripping of the adsorbed organics and subsequent incineration. Handling and energy costs are high, making carbon treatment of leachate cost effective only for removal of residual organics. However, carbon adsorption may appear attractive in special cases because it is simple, effective, and requires low capital. In addition, mobile units are available for emergency or temporary treatment systems.

An interesting application of carbon adsorption is zimpro's Powdered Activated Carbon Technology (PACT). Zimpro claims that when the powdered carbon is added to a biological contactor, the carbon adsorbs the toxic organics which would normally inhibit microbe growth. The toxics are slowly de-adsorbed and safely consumed. Zimpro recommends PACT for:

* wastewater which has frequent shocks in COD loading;
* hard-to-treat industrial wastes; and
* wastes requiring nitrification.

All the problems above are typically associated with leachate treatment.

CONCLUSION

There is much discussion of moving away from landfills and toward incineration, recycling, waste to energy, or other methods of garbage disposal. Realistically though, mankind, particularly in developed countries such as our own, constantly produces more waste, and must develop new disposal sites to manage it. Even incinerators and waste to energy plants produce tons of ash that must be disposed of in an environmentally sound manner, and recycling can handle only 50-70% of a community's solid waste. It is obvious that operating landfills will be around for a long time, as well as the thousands of closed landfill sites that may require environmental remediation.

Management of leachate from sanitary landfills is sure to be a significant concern in coming months and years. EPA is scheduled to release proposed regulations on nonhazardous waste landfills under Subtitle D of the Resource Conservation and Recovery Act in 1988. These regulations are expected to have very specific requirements for liners, leachate collection systems, groundwater monitoring requirements, final caps, and financial assurance for post-closure maintenance. State and local regulations are also becoming much more stringent.

Landfill designers and operators will be challenged to minimize leachate generation. Wastewater equipment manufacturers will be challenged to design more efficient treatment processes for complex leachates. The future will require the close cooperation of regulators, engineers, and landfill owner/operators to ensure protection of national resources in a fiscally and environmentally responsible manner.

TABLE 1
TYPICAL LEACHATE COMPOSITIONS

Landfills Parameter (mg/L)	BFI Ranges
COD	230-50,000
BOD_5	170-14,000
TOC	N/A
pH	5.9-7.5
TDS	590-15,000
TSS	170-600
Specific Conductance	1,000-8,000
Alkalinity (as $CaCO_3$)	115-2,500
Hardness (as $CaCO_3$)	N/A
Total P	N/A
Ortho P	N/A
NH_4-N	N/A
NO_3+NO_2-N	2-1,250
Ca	N/A
Cl	N/A
Na	N/A
K	N/A
SO_4	N/A
Mn	N/A
Mg	N/A
Fe	0.05-340
Zn	0.01-15
Cu	0-1.3
Cd	0-.01
Pb	0-1.5

Notes:
N/A - Data Not Available

TABLE 2
RECOMMENDED LEACHATE CHARACTERIZATION PARAMETERS

BOD_5: 5 Day Chemical Oxygen Demand. A measure of oxygen consumed in 5 days by inoculated microbes.

COD: Chemical Oxygen Demand. Oxygen consumed via acidified potassium dichromate. Directly related to TOC analysis. Comparison with BOD_5 can indicate biodegradability of leachate.

TOC: Total Organic Carbon. An accurate measure of total carbon.

pH: Hydrogen activity. Neutralization required if pH out of range. pH value may reflect leachate age. Take care when sampling; exposure to air can cause change.

TDS: Total Dissolved Solids. Indicates total dissolved salts. High values can affect biodegradability. Specific conductance and TDS usually correlated.

TSS: Total Suspended Solids. Suspended solids are usually organic and impact COD demands. High TSS indicates need for primary clarification.

Ortho-P: Total ortho phosphorous utilized by microbes for cell growth. Phosphorous addition may be necessary.

NH_4-N: Ammoniacal Nitrogen. Excess concentration may inhibit microbe growth. Need for nitrification (conversion to oxidized nitrogen) may control sizing of aerobic treatment units.

NO_3-NO_2-N: Oxidized nitrogen as nitrate and nitrite anions. Contributes to eventual nitrogen effluent loadings.

Heavy Metals: Fe, Zn, Cu, Cd, Pb, Cr (III & IV), Hg, Ni, Ag. Excessive concentrations of heavy metals inhibits microbe growth. Local authorities may impose maximum concentration limits in effluent.

Organic Chemicals Scans: Alcohols, Aliphatics, Amines, Aromatics, Halocarbons, Pesticides, Phenols, Phthalates. High concentrations may be toxic to microbes. Analysis necessary only if required by authorities or leachate is suspect. A lot of data available on inhibiting effects (see Ref. 10).

Coliforms: Indicates pathogen contamination. Leachate may require disinfection before discharge.

LITERATURE CITED

1. Bouwer, Herman. Groundwater Hydrology, McGraw-Hill, Inc. New York. 1978.

2. Ehrig, H.J., Waste Management and Research, pp. 53-68. Vol. 1, 1983.

3. Kennedy, M.O., York, R.J. Proceedings of the National Leachate/Liner Conference. St. Louis, MO, 1986.

4. Lu, J., et. al., Leachate from Municipal Landfills, Noyes Publications, New Jersey, 1985.

5. Metcalf & Eddy, Inc. Wastewater Engineering. Second Ed. McGraw-Hill Book Company. New York. 1979.

6. Mitchell, R., ed. Water Pollution Technology. Vol. 2. Wiley-Interscience, 1978.

7. Peters, R.W., Ku, Y. AIChE Symposium Series. 243, Vol. 81, 1985.

8. Salimando, J., Waste Age, p. 28, November, 1987.

9. Sanitary Landfill Gas and Leachate Management, Course Notes. University of Wisconsin-Madison, December 1986.

10. Shuctrow, A.J., Pajak, A.P., Touhill, C.J. U.S.E.P.A. Report No. SW-871. Cincinnati, OH, 1980.

11. Steiner, R.L., Keenan, J.D., Fungarolli, A.A. U.S.E.P.A. Report No. SW-758. Washington, D.C., 1979.

12. Woelfel, G.C., Schaber, P.E., Carter, J.L. Proceedings of the National Leachate/Liner Conference. St. Louis, MO, 1986.

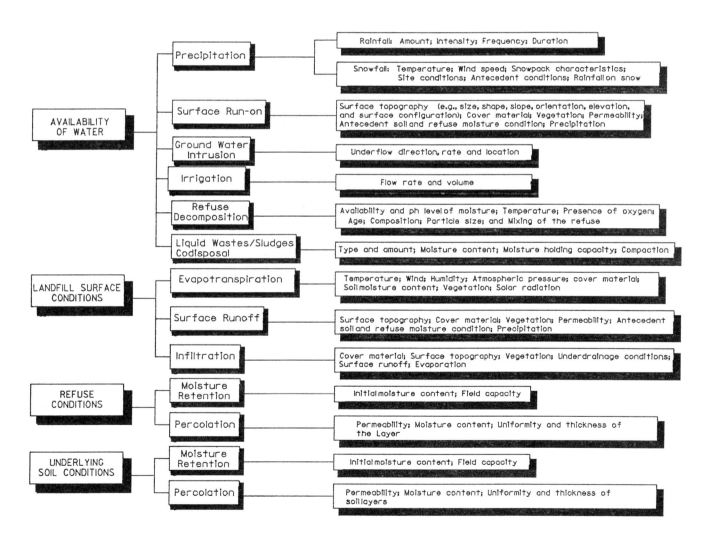

Figure 1. Factors affecting leachate volume generation.
from: Leachate from municipal land fills, Lu., et.al., 1985.

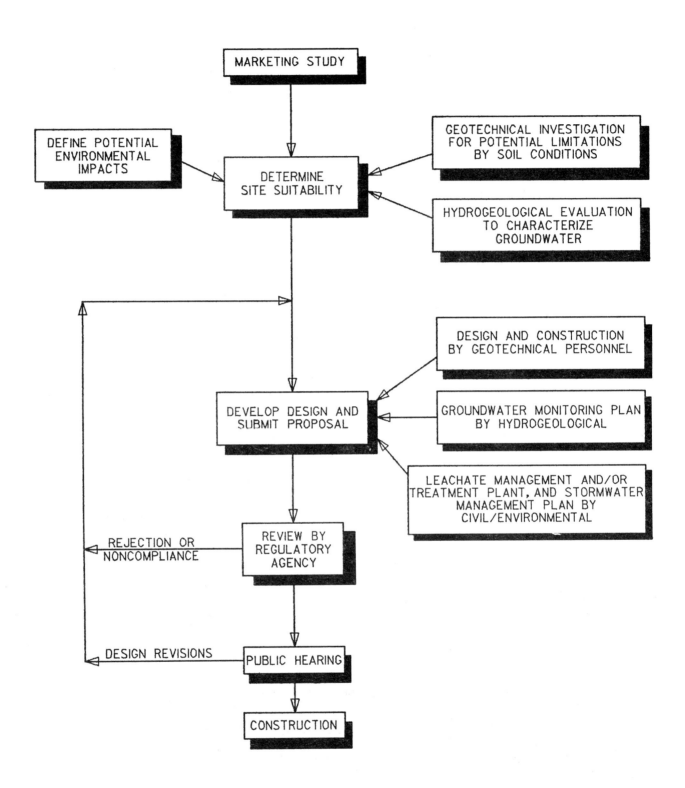

Figure 2. Landfill design and permitting processes.

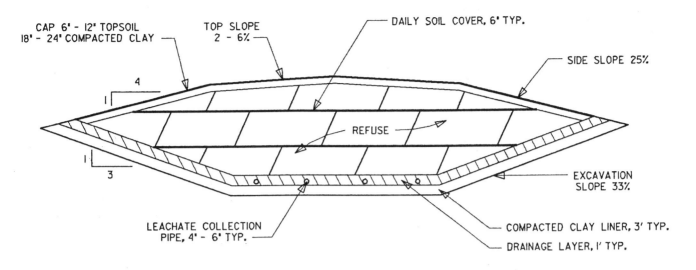

Figure 3. Cross section of sanitary landfill.

Figure 4. Schematic of an operating sanitary landfill.

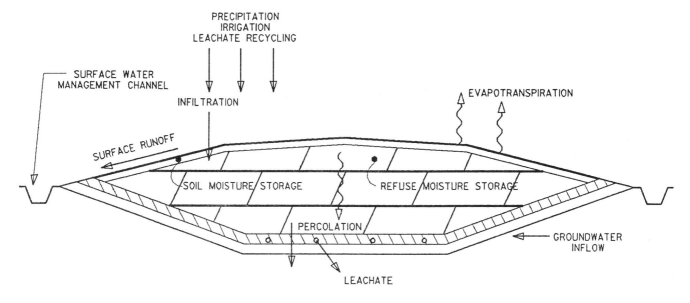

Figure 5. Components of water balance for sanitary landfill.

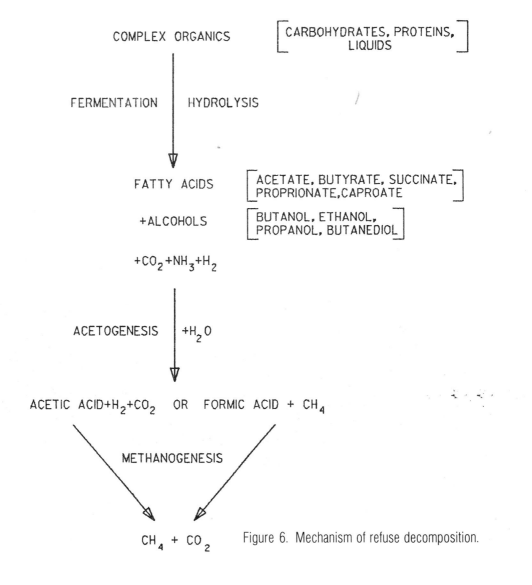

Figure 6. Mechanism of refuse decomposition.

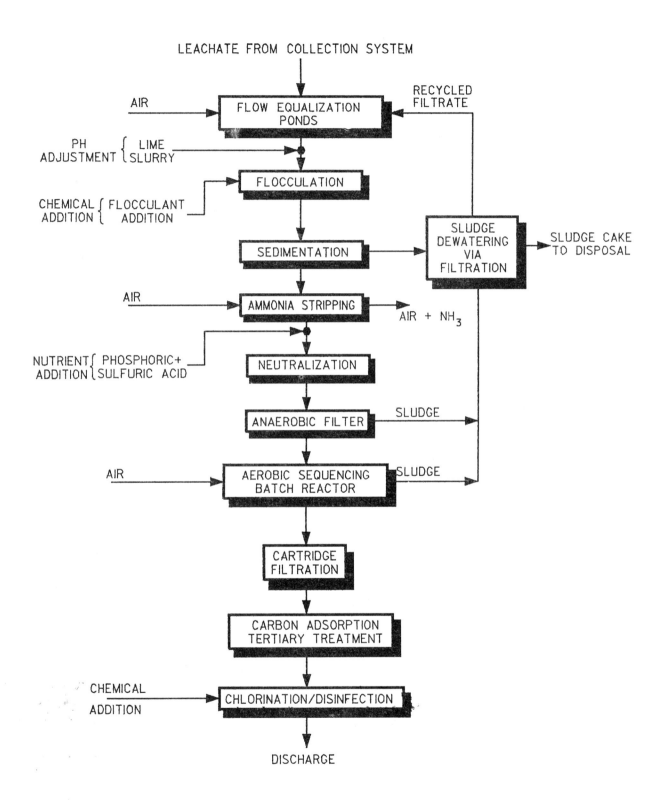

Figure 7. Hypothetical leachate treatment system process flow diagram.

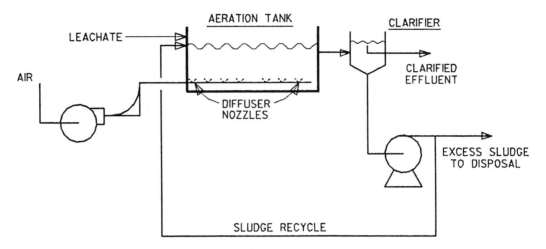

Figure 8. Activated sludge process.

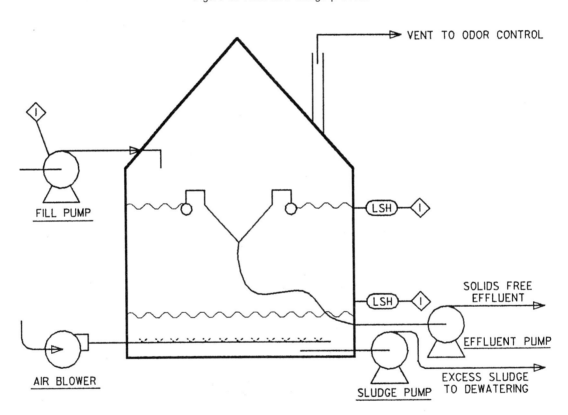

SEQUENCE STEPS

① TANK FILL TO LSH WITH BLOWER ON.
② SHUTDOWN FILL PUMP.
③ AERATE FILL TANK FOR SPECIFIED TIME.
④ SHUTDOWN BLOWER.
⑤ ALLOW SLUDGE SETTLING TIME.
⑥ START EFFLUENT PUMP: DECANT SUPERNATANT TO LSH.
⑦ SHUTDOWN EFFLUENT PUMP.
⑧ PUMP OUT EXCESS SLUDGE FOR SPECIFIED TIME.
⑨ END OF SEQUENCE.

Figure 9. Sequencing batch reactor.

THE USE OF FLEXIBLE MEMBRANES IN PROTECTION OF GROUNDWATER

Robert E. Landreth ■ Risk Reduction Engineering Laboratory*, United States Environmental Protection Agency, Cincinnati, OH 45268

Flexible membrane liners are increasingly being used to prevent leachate and other waste fluids generated at landfills, surface impoundments, and waste piles from entering the groundwater. The Hazardous and Solid Waste Amendments (HSWA) of 1984 requires liners and leachate collection components to be chemically resistant to the fluids to be contained, be constructed of materials that can be properly installed, maintained and repaired, and be durable.

This paper will describe the EPA's research program on treatment, storage and disposal facilities with emphasis on containment systems. The paper will identify landfill construction concerns and potential solutions to those concerns.

INTRODUCTION

Land disposal continues to be a popular and economical method of disposing of hazardous wastes. Though poor practices in the past have led to significant environmental problems, today's properly designed, constructed, and operated land disposal facility can ensure that wastes will not be a source of health and environmental concern. The Hazardous and Solid Waste Amendments (HSWA) of 1984 requires liners and leachate collection components for treatment, storage and disposal facilities (TSDF).

Improved containment technology has been the primary thrust of the U.S. Environmental Protection Agency's Land Disposal Research Program. The technology universally includes liners to prevent liquid (leachate) escape, systems to collect and remove liquid and gas whose presence cannot be prevented, and covers to exclude infiltration and control gas emissions.

CONTAINMENT TECHNOLOGY

Liners

To protect human health and the environment, certain landfills and surface impoundments are required, under Sections 3004(o) and 3015 of the HSWA, to have two or more liners and a leachate collection system above and between such liners. One design (1,2,3) the Agency believes meets these requirements for landfills is depicted in Figures 1 and 2.

This liner system includes a top liner and a composite bottom liner. The top liner is designed and constructed and then operated to prevent the migration of any hazardous constituents into the bottom liner during the period the facility remains in operation. Flexible membrane liners (FML) appear to be the only practical available material that will prevent migration. The bottom liner consists of two components that are intended to function as one system, hence, the term "composite" liner. The composite bottom liner is the best technology to facilitate rapid detection of any leakage of leachate through the top liner, enhance the efficiency of leachate collection, and minimize leachate migration from the unit. Like the top liner, the upper component of the bottom liner, also a FML, is designed, constructed, and operated to prevent the migration of any constitutent into the lower component during the period of facility operation. The lower component of the bottom liner is designed, constructed, and operated to minimize the migration of any

---

*The name of the Laboratory has been renamed from Hazardous Waste Engineering Research Laboratory to Risk Reduction Engineering Laboratory.

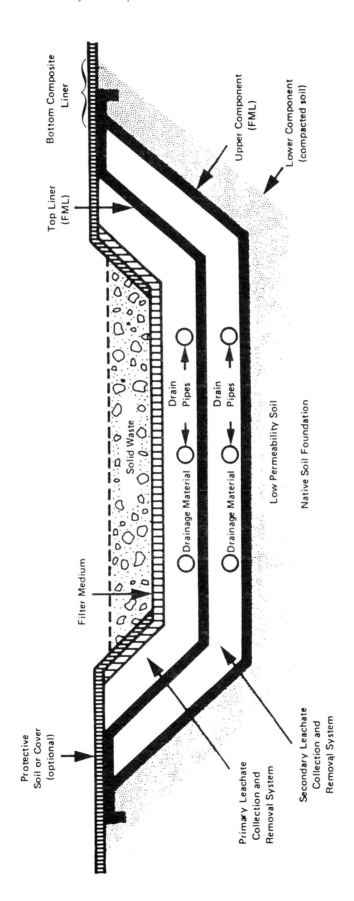

FIGURE 1
SCHEMATIC OF AN FML/COMPOSITE DOUBLE LINER SYSTEM
FOR A LANDFILL

FIGURE 2
SCHEMATIC PROFILE OF AN FML/COMPOSITE
DOUBLE LINER SYSTEM FOR A LANDFILL

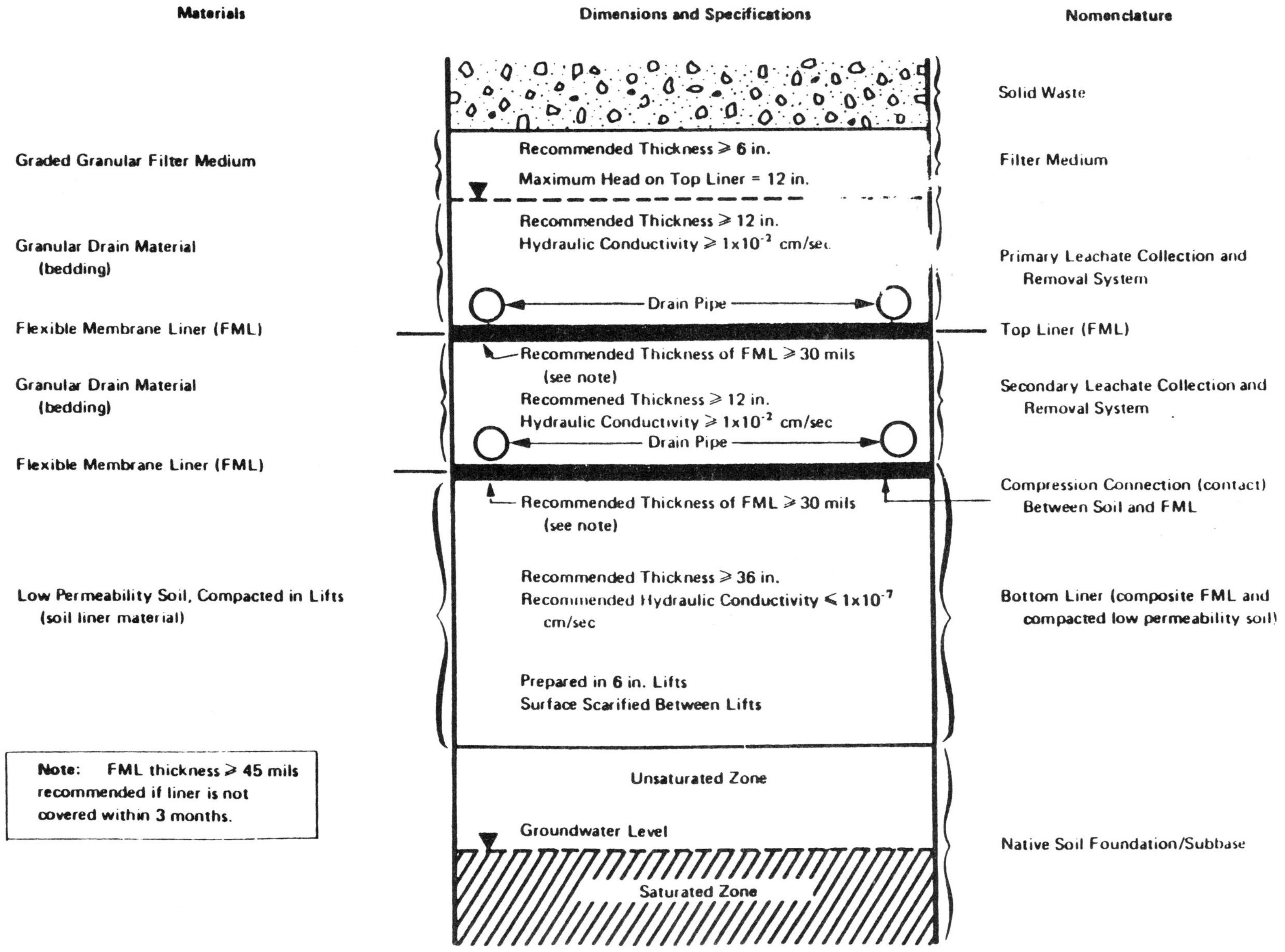

Materials

- Graded Granular Filter Medium
- Granular Drain Material (bedding)
- Flexible Membrane Liner (FML)
- Granular Drain Material (bedding)
- Flexible Membrane Liner (FML)
- Low Permeability Soil, Compacted in Lifts (soil liner material)

Dimensions and Specifications

- Recommended Thickness ≥ 6 in.
- Maximum Head on Top Liner = 12 in.
- Recommended Thickness ≥ 12 in.
- Hydraulic Conductivity ≥ 1×10^{-2} cm/sec
- Drain Pipe
- Recommended Thickness of FML ≥ 30 mils (see note)
- Recommended Thickness ≥ 12 in.
- Hydraulic Conductivity ≥ 1×10^{-2} cm/sec
- Drain Pipe
- Recommended Thickness of FML ≥ 30 mils (see note)
- Recommended Thickness ≥ 36 in.
- Recommended Hydraulic Conductivity ≤ 1×10^{-7} cm/sec
- Prepared in 6 in. Lifts
- Surface Scarified Between Lifts
- Unsaturated Zone
- Groundwater Level
- Saturated Zone

Nomenclature

- Solid Waste
- Filter Medium
- Primary Leachate Collection and Removal System
- Top Liner (FML)
- Secondary Leachate Collection and Removal System
- Compression Connection (contact) Between Soil and FML
- Bottom Liner (composite FML and compacted low permeability soil)
- Native Soil Foundation/Subbase

Note: FML thickness ≥ 45 mils recommended if liner is not covered within 3 months.

constituent through the bottom component if a breach in the upper component were to occur before the end of facility operation. The lower component of the composite liner is a compacted soil that should meet the technical requirements identified in Figure 2.

Leachate Collection and Removal Systems

The primary and secondary leachate collection and removal systems (Figure 2) should: be designed to ensure that the leachate depth above the top liner does not exceed 1 foot; be constructed of materials that can withstand the chemical attack that results from waste liquids or leachates; be designed and constructed to withstand the stresses and disturbances from overlying wastes, waste cover materials, and equipment operation; and be designed and operated to function without clogging through the post-closure monitoring period. Components should be properly installed to ensure that the specified performance of the leachate collection system is achieved.

The primary leachate collection and removal system should have:

(a) At least a 30-cm (12-inch) thick granular drainage layer that is chemically resistant to the waste and leachate, with a hydraulic conductivity not less than 1×10^{-2} cm/sec, 1 cm/sec for the secondary leachate collection system, and with a minimum bottom slope of 2 percent. Innovative leachate collection systems incorporating synthetic drainage layers or nets may be used if they are shown to be equivalent to or more effective than the granular design with respect to chemical compatibility, flow under load, and protection (e.g., from puncture) of the FML.

(b) A graded granular or synthetic fabric filter above the drainage layer to prevent clogging. Criteria for graded granular filters and for synthetic fabric filters are found in numerous publications such as the Geotextile Engineering Manual available from the Federal Highway Administration. The granular drainage material should be washed to remove fines before installation.

(c) A drainage system of appropriate pipe size and spacing on the bottom of the unit to efficiently collect leachate. These pipe materials should be chemically resistant to the waste and leachate. The piping system should be strong enough to withstand the weight of the waste materials placed on it and vehicular traffic operated on top of it.

(d) A primary leachate collection system that covers the bottom and sidewalls of the units.

(e) A sump in each unit or cell capable of automatic and continuous functioning. The sump should contain a conveyance system for removing leachate from the unit, e.g., either a sump pump and conveyance pipe or gravity drains.

COVERS

The cover system is a multilayer system designed to minimize the infiltration of precipitation into the waste management facility after closure. To prevent the "bath tub" effect, the cover system must be no more permeable than the liner system. It must operate with minimum maintenance and promote drainage from its surface while minimizing erosion. It must also be designed so that settling and subsidence are accommodated to minimize the potential for destruction of continuity or function of the final cover.

As a minimum, the cover consists of the vegetation, drainage, and barrier layers (Figure 3). While the dimensions shown in Figure 3 for the various layers are recommended minimum values, the cover design may be modified based on the location of the unit. The vegetation layer should be sufficiently thick and of a quality to support vegetation. The vegetation provides an aesthetically pleasing surface, minimizes erosion potential, and assists in removing excess moisture through evapotranspiration. The final top slope, after allowance for settling and subsidence, should be between 3 and 5 percent.

The drainage layer may consist of a geotextile to prevent soil from clogging the sand drain and the sand drain itself. Recent designs using geosynthetic composites have been suggested. The permeability of the sand drain should be equal to or greater than 10^{-3} cm/sec and have a final bottom slope of at least 2 percent.

The barrier layer should have two components: a flexible membrane liner and a recompacted soil layer. The membrane should be protected on both sides by a bedding layer, be a minimum of 20 mil, have a final slope of at

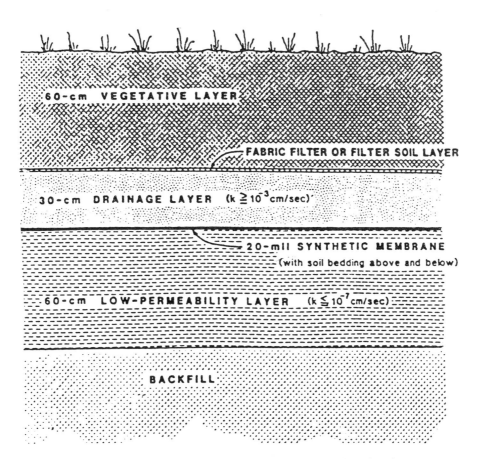

Figure 3. Minimal cover components as described in RCRA guidance.

least 2 percent and be located wholly below the average depth of frost. The recompacted soil layer should have a saturated hydraulic conductivity of not more than 10^{-7} cm/sec and be compacted in lifts.

Other site specific considerations may be a gas collection and vent system, a barrier for burrowing animals, and a barrier to prevent deep rooted vegetation from clogging the drainage layer.

The design and construction of cover systems is not unlike that for the liner systems except in selecting the membrane. Although the membrane may be exposed to gases, experience has shown that, in general, diffusion of gas through membranes has not been detrimental. If this is a concern, then the appropriate test should be performed.

Long-term maintenance is probably the major concern with cover systems. A systematic maintenance plan should be developed to ensure that the cover system will maintain its integrity.

CONSTRUCTION CONCERNS

A number of questions have arisen regarding how to design the individual components, how to determine their chemical resistance, and what type of quality assurance plan is necessary to ensure that actual construction meets or exceeds the design requirements and therefore, complies with the above guidance on liner and leachate collection and removal systems.

Design of Components

In the past, designs using FML and other synthetic materials have been based primarily on what was available from the liner industry and on the experience of the design firm. Although synthetic materials were never meant to be nor are they now considered to be structural components within a waste management facility, civil engineers have been developing designs based on free-body diagrams. The designs employ the physical as well as the hydraulic properties of the synthetic materials. The three broad classes of synthetic materials considered in the designs are FML used as hydraulic barriers, geotextile intended for applications within soils, and geogrids/geonets used to form leachate collection and removal systems. Combinations of these components are used to enhance the overall performance of the systems.

A recent manual (4) by the Agency evaluates designs of components beneath, within, and above a waste management unit, cons tion/fabrication considerations and long-term performance considerations. The designs are based on free-body equilibrium equations. A design ratio (ratio of the allowable material performance divided by the actual material service conditions calculated) is used for designs where clear limits are known. This approach, unfortunately, has very few actual field data to verify the accuracy of the solutions, and it lacks standardized test procedures, especially for the geotextiles and geonets/geogrids. Research is currently underway to provide the field verification. The manual is useful in identifying potentially weak areas within a system design. Using a conservative design approach should prevent the failure of the system for this specific design function.

Chemical Resistance of Components

The chemical resistance of all the system components that come in contact with leachate is a major concern. The Agency-developed Method 9090 defines the basic framework for determining whether the component materials are chemically resistant. After exposing samples of the component to representative samples of the leachate at room and elevated temperatures, a component's physical properties are determined at 0, 30, 60, 90, and 120 days. In evaluating these data, a rate of change is being sought for the various exposure times. A continuing change suggests no chemical resistance, whereas no rate of change suggests chemical resistance. The gray area between these extremes requires expert interpretation. The Agency has developed computer programs that simulate the advice of experts by incorporating their decision, rules or criteria in terms of "IF---THEN" statements. These programs, called expert systems, are essentially large pattern matching routines that seek a solution that corresponds to the pattern of input data. These expert systems offer significant opportunities for ensuring expedient implementation of the latest technologies, compliance with regulatory policies, and consistency in design decisions.

Quality Assurance/Quality Control

The third major concern in the development of a waste management facility is how to construct a facility that meets or exceeds the design specifications. Recent EPA studies (5) concluded that the most common fault in land disposal systems leading to failure lies in construction deficiencies in the containment system. These studies further suggest that a construction program involving a good quality assurance and quality control (QA/QC) plan could dramatically reduce disposal system failures. With respect to waste management facilities, EPA defines construction quality management as a process that employs scientific and engineering principles and practices to ensure, with a reasonable degree of certainty, that a constructed waste management facility meets or exceeds the specified design. Construction quality management consists of construction quality assurance and construction quality control. Construction quality assurance (CQA) involves a planned system of overview activities to ensure that construction quality is being effectively implemented: that the adequacy and effectiveness of the QA/QC plan is continuously evaluated, that QA/QC activities are carried out, that the data are collected, and then interpreted in accordance with the plan. Construction quality control (CQC) involves a planned system of inspection activities that are used to directly monitor and control the quality of a construction project. QA/QC plans are written documents identifying areas of responsibility and lines of authority; qualification of personnel; inspection activities; sampling strategies including frequency and acceptance/rejection criteria; and documentation procedures.

SUMMARY

Landfill disposal facilities that are properly designed, constructed, and operated continue to provide an economical and safe method for disposing of hazardous waste. Landfills with two (or more) FML liners, a primary and secondary leachate collection and removal system that can withstand chemical attack, and an aesthetically pleasing cover system with adequate vegetation, drainage, barrier layer can meet governmental regulations.

EPA has and continues to support research into specific areas of concern, e.g., the design and chemical resistance of a land disposal system's component parts and the establishment of guidelines for QA/QC plans relating to the construction of a land disposal facility.

REFERENCES

1. USEPA (1985). Guidance on Implementation of the Minimum Technological Requirements of HSWA of 1984, Respecting Liners and Leachate Collection Systems. Reauthorization Statutory Interpretation #5. Washington, D.C.: U.S. Environmental Protection Agency. USEPA 531/SW-85-012. 1985.

2. USEPA (1985). Minimum Technology Guidance on Single Liner Systems for Landfills, Surface Impoundments and Waste Piles - Design, Construction and Operation. Washington, D.C.: U.S. Environmental Protection Agency. USEPA 530/SW-85-013. 1985.

3. USEPA (1985). Minimum Technology Guidance on Double Liner Systems for Landfills - Design, Construction and Operation. Washington, D.C.: U.S. Environmental Protection Agency. USEPA 530/SW-85-014. 1985.

4. G. N. Richardson and R. M. Koerner (USEPA). Geosynthetic Design Guidance for Hazardous Waste Landfill Cells and Surface Impoundments. Cincinnati, OH. PB 88-131263. Also available from the Geosynthetic Research Institute, Drexel University, Philadelphia, PA.

5. USEPA (1986) Technical Guidance Document: Construction Quality Assurance for Hazardous Waste Land Disposal Facilities. Cincinnati, OH. EPA/530/SW-86-031.

A CASE STUDY OF SANITARY LANDFILL LEACHATE ATTENUATION BY A NATURAL CLAY-PEAT SOIL IN THE GULF COAST REGION

Michael J. Brady ■ Geoscience, Browning-Ferris Industries, P.O. Box 3151, Houston, TX 77253

This paper discusses the rationale, design, implementation, and findings of a field and laboratory investigation of the chemical and biochemical attenuation of landfill leachate by a natural clay-peat soil to assist in evaluating the engineering design of site closure plans.

BACKGROUND

Browning-Ferris Inc. acquired a company in late 1972 which operated a sanitary landfill as part of its activities in a gulf coast municipality. The landfill had been in operation since 1962 before BFI acquired the site. Operations at the landfill were upgraded, and in 1985 the site was reviewed for expansion potential. Part of this internal review involved an evaluation of the impact of the existing landfill on ground water to help evaluate necessary design considerations for an expansion as well as the need to consider upgrading containment around the older portions of the landfill.

SITE DESCRIPTION

The landfill is located in a low lying area, surrounded on three sides by shallow brackish water. A dike was constructed around the landfill to a height of approximately 10 feet, and the landfill rises to a maximum height of 20 feet above the water level. A municipal water treatment plant is adjacent to the site and discharges treated effluent to the surrounding water.

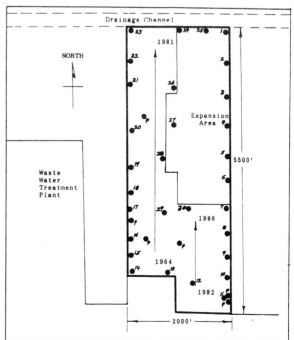

GENERALIZED SITE PLAN

Showing Locations of Soil Borings and Piezometers

SITE GEOLOGY

In order to characterize the geology and ground water conditions at the site over 30 soil borings and piezometers have been installed, and the piezometers used for sampling water quality and water levels. A generalized geologic column illustrates the stratigraphic sequence.

The site geology is unique. The uppermost stratigraphic unit is a layer of peat and soft clay, ranging from 6 to 16 feet in thickness. Access for drilling equipment has been obtained in filled areas, so boring logs show the solid waste fill placed over this uppermost natural unit. Below this is a very soft to soft, highly plastic clay with interspersed silt lenses which extends down an average of 45 feet below the peat layer. Below this clay is a 3 - 5 foot thick sand layer which is underlain by older soft to medium clays of Pleistocene age.

No shallow potable water quality aquifers were found beneath the site, and the uppermost known potable water bearing sand is about 600 feet deep, having a probable brackish nature at this location.

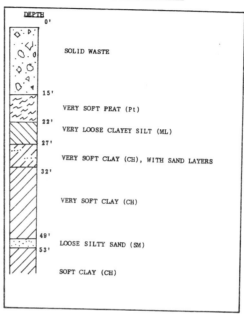

GENERALIZED STRATIGRAPHIC COLUMN

SITE HYDROGEOLOGY

Site hydrogeology has been analyzed at the site in several studies.
Ground water flow has been characterized in two zones:
1. The shallow (water-table) zone (aquifer), extending downward from surface to approximately 27 feet, consisting of Solid Waste, Peat, and the Silty Clay/Clayey Silt layers above the bounding clay layer from 27 to 49 feet.
2. The deep zone, consisting of a four foot thick layer of loose silty sand.

The ground water flow in the shallow zone appeared to be irregularly radial, elongated in the north-south direction, and strongly influenced by site topography.

Flow in the lower zone is southeasterly across the site.

Ground water quality at the site is moderately saline in both of the zones monitored, and not potable by drinking water standards.

Chloride levels range from 2800 - 4200 milligrams/liter, and total dissolved solids range from 5800 to 8800 milligrams/liter. Twenty-one soil borings and nineteen piezometers were installed for the major hydrogeologic study at this site, and a number of the piezometers were regularly sampled as monitoring points for ground water quality.

Permeabilities were determined by a series of falling head and static head field tests, as well as laboratory tests. The shallow zone had an average horizontal permeability of 1.5×10^{-5} centimeters per second (cm/sec). The clay below the shallow zone showed an average horizontal permeability of 2.2×10^{-4} cm/sec and an average vertical permeability of 1.0×10^{-7} cm/sec.

Later tests of the permeability of samples obtained beneath the landfill which had been compressed by the weight of the landfill or compacted by operating equipment demonstrated permeabilities less than 10^{-9} cm/sec.

SCOPE OF RECENT SITE ASSESSMENT PROGRAM

Evaluation of the landfill site to determine potential for contaminant migration by lateral flow through the upper peat stratum into the surrounding brackish waters became of concern for the design of the final expansion of the landfill, and design of site containment, if necessary, for post closure care. Several engineering alternatives had been examined, and selection of the most suitable alternative prompted this in-depth study.

R. S. Reimers, A. A. Abdelghani, and V. C. Culpepper of Tulane University School of Public Health and Tropical Medicine, Department of Environmental Health Sciences, in co-operation with Soil Testing Engineers, Inc. of Baton Rouge, Louisiana proposed the following scope of work:

1. Testing and analysis of samples of soil/peat, leachate, and ground water to determine the attenuation/retardation coefficients, etc. for ground water analysis.

2. Geochemical analysis to determine probable changes in leachate/ground water chemistry due to flow through the peat fraction.

3. Reporting of results to assist BFI in determining potential impacts of engineering design alternatives, assessing the present site conditions, and developing engineering plans to minimize environmental impact of the landfill.

RATIONALE FOR THE TESTING PROGRAM

A literature review was conducted and the results tabulated to develop a rationale for selecting parameters and tests to evaluate site conditions and potential for contaminant migration. Properties familiar to the researchers were used to select parameters to be studied. Copper, cadmium, thallium and lead were selected as typical representative metals. Studies of PCB's and PCP distributions in nature had been reviewed and were available as guidance for examining priority pollutant organics in this study. A set of organic compounds were selected to represent volatile, base neutral, and acid extractable organics.

The research project was planned for three phases:

1. Sample collection; leachate and ground water characterization.
2. Batch laboratory testing and analysis to determine attenuation and degradation of organics and metals, and the capacity of the peat for attenuation.
3. Data analysis and report preparation.

A fourth phase was added later.

In Phase 1, leachate and ground water samples were collected and analyzed to estimate the attenuation by peat of leachate constituents at the site. Four samples of leachate and four samples of ground water were analyzed for priority pollutants and gross indicator parameters.

LEACHATE AND GROUNDWATER ANALYSES		
CONSTITUENT	FREQUENCY	METHOD SOURCE
PRIORITY POLLUTANTS*	8	STANDARD METHODS
EIRA	8	" "
PH	8	" "
SP. CONDUCTIVITY	8	" "
TOC	8	" "
BOD_5	8	" "
COD	8	" "
TDS	8	" "
TEMPERATURE	8	" "
ORP	8	" "
ALKALINITY	8	" "
HARDNESS	8	" "
D.O.	8	" "

* THESE ANALYSES WERE FOR BACKGROUND ASSESSMENT AND ESTIMATED TESTING REQUIREMENTS. ALL OTHER ANALYSES WERE TO MEASURE SOIL INFLUENCE.

Phase 2 testing involved the development of batch adsorption isotherms on uncontaminated and leachate-exposed peat with selected compounds to determine the maximum capacity for peat to bind the compounds.

PEAT CHARACTERIZATION ANALYSES			
CONSTITUENT	FREQUENCY	COMMENTS	METHOD SOURCE
Organic Content	6	Ability to Adsorb Organics	Std. Soil Analysis
pH	6	Ionization Influence	"
Total Phosphorus	6	Nutrient Level	"
Total Nitrogen	6	" "	"
Total Sulfide	6	Soil Functional Content	"
Totak K. Nitrogen	6	" " "	"
Moisture Content	6	Bacterial Activity	"
Cation Exchange	6	Metals Capacity	"
Conductivity	6	Salinity-Salts Content	"
Sodium Adsorption Ratio	6	Hardness to Na Content	"
Particulate Size and Fraction	6	Ability to Sorb Toxics	"
VSS and SS	6	Potential Biological Activity	Std. Methods
BOD5 and COD	6	" " "	" "

Column percolation tests were also conducted to test mass transfer rates at different flow rates. The Weber approach was used instead of the Bohart-Adams approach to reduce testing time from three months to one week.

RESULTS OF THE TESTING PROGRAM

1. Zinc, nickel, copper, and mercury are sorbed in the peat zone, some to below detection limits.
2. Selenium, chromium, arsenic, and beryllium were not detected or were observed below drinking water standards in the ground water, and were weakly attenuated or not attenuated by the peat zone.
3. Silver was also reduced in the peat zone and not observed above drinking water limits in the ground water.
4. Three priority pollutant organics were detected in leachate samples but not in ground water. Only methylene chloride was detected in both the leachate and the ground water.
5. Heavier aromatic volatile organics appeared to be sorbed and drop below detectable limits.
6. No other volatile, base neutral, or acid extractable organics were detected or identified in the leachate and in ground water.
7. BOD_5 was reduced to approximately 10 mg/l, regardless of influent concentrations ranging from 24 to 1490 mg/l.
8. Organic COD's were reduced by 60% to 90% depending on initial COD influent concentrations.
9. Both degradable and non-biodegradable TOC's were readily removed by the peat to a level of around 35 mg/l.
10. Phenols and cyanide levels were reduced to concentrations below drinking water standards.

INFLUENT AND EFFLUENT COLUMN ANALYSES		
CONSTITUENT	FREQUENCY	COMMENTS
pH	8	Note change due to peat
TOC	8	Look at change in organics
COD	8	Look at column redox activity
Conductivity	8	Est. influence of activity
ORP	8	Look at column redox activity
PERC	8	Examine breakdown products
Trichlorobenzene	8	Examine these and phenols for breakdown products
Toluene	8	Examine breakdown products
Thallium	8	Examine capacity & conversion
Cadmium	8	" " " "
Lead	8	" " " "
Copper	8	" " " "

Samples to be taken on days 0, 1, 2, and 3.

BATCH TEST ANALYSES		
CONSTITUENT	FREQUENCY	COMMENTS
pH	8	Note equilibrium
ORP	8	" "
TOC	8	" "
Conductivity	8	" "
Temperature	8	For adsorption isotherms
PERC	8	Capacity & breakdown prod.
Trichlorobenzene & Phenols	8	" " " "
Toluene	8	" " " "
Thallium	8	Capacity & conversion
Cadmium	8	" " "
Lead	8	" " "
Copper	8	" " "

Additionally, the sorption data indicated that the removal mechanism was not due to adsorption except for toluene. This indicates that clays in the peat zone, organic complexes, and biological activity may be principal attenuating mechanisms. Cadmium, thallium, and lead showed the least peat adsorption of the metals, but the concentrations of these metals were near U. S. EPA Drinking Water Standards, and lower than ambient water concentrations.

Based on observations during sampling, a fourth phase of surface water sampling was added. This provided a surprising and significant set of results.

The Phase 4 program was conducted in the surface water surrounding the site. Composite samples were taken at the surface, at the bottom, and at mid water depth along four transects radiating from the landfill. Samples were taken 10, 30, 50, 100, and 200 feet from the edge of the landfill. Results of this program were surprising.

```
          SURFACE WATER DATA - AVERAGE VALUES FOR 4 TRANSECTS
          NORMALIZED TO NEAREST SAMPLING POINT TO LANDFILL

                    DISTANCE FROM LANDFILL
PARAMETER      10'    30'    50'    100'   200'  OBSERVATIONS

pH            1.00   0.91   0.86   0.87   0.85   DECREASE AWAY
D.O.          1.00   0.99   0.94   0.86   0.80   DECREASE AWAY
ORP           1.00   0.90   0.96   0.97   1.00   ?

SP. COND.     1.00   0.96   0.98   1.00   1.02   3 OF 4 INCR. AWAY
CHLORIDES     1.00   1.00   1.04   1.27   1.47   INCREASE AWAY
TDS           1.00   1.11   1.04   1.14   1.08   3 OF 4 INCR. AWAY

MERCURY       1.00   0.86   1.20   1.05   1.10   INCREASE AWAY
CADMIUM       1.00   1.21   1.18   1.11   1.08   INCREASE AWAY
LEAD          1.00   1.06   1.06   1.04   1.18   INCREASE AWAY
IRON          1.00   0.80   1.12   1.39   0.72   DECREASE AWAY
MANGANESE     1.00   0.84   1.94   1.18   1.68   INCREASE AWAY

BOD           1.00   0.99   1.65   2.73   1.11   INCREASE AWAY
COD           1.00   1.08   1.44   0.93   0.85   INCREASE AWAY
TOC           1.00   0.96   0.95   1.52   0.93   ?
```

Chlorides, TDS, Iron, and Manganese were used as indicators of leachate impact, and did show some degree of site impact. Phenols and Cyanide, which did not show much attenuation in the lab, did not indicate leachate was a major source to surface water.

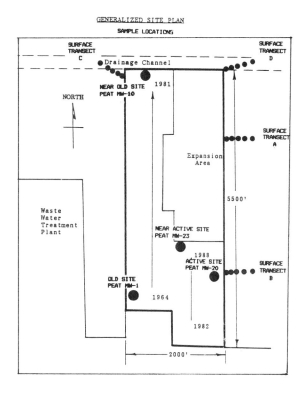

GENERALIZED SITE PLAN
SAMPLE LOCATIONS

```
              FIELD SAMPLING DATA
         LEACHATE & GROUND WATER ANALYSES

1. 52 ORGANIC COMPOUNDS WERE DETECTED IN LEACHATE
   WHICH WERE NOT DETECTED IN GROUND WATER SAMPLES.

2. 13 ORGANIC COMPOUNDS WERE DETECTED IN GROUND WATER SAMPLES
   WHICH WERE NOT DETECTED IN LEACHATE SAMPLES.
   THESE MAY INDICATE A DIFFERENT SOURCE OF CONTAMINANTS.

3. 5 SEMI-VOLATILE COMPOUNDS INDICATED 37% - 63% REMOVAL.

4. METHYLENE CHLORIDE, CHLOROBENZENE, TOLUENE, AND ETHYL BENZENE
   SHOWED REMOVAL OF 33% TO 98% OF THE COMPOUNDS.

5. INDICATED METALS REMOVAL IS TABULATED BELOW:

                    INDICATED
METAL          PERCENT REMOVED

ANTIMONY          0 - 98
ARSENIC           0 - 93
BARIUM            0 - 73
BERYLLIUM         0 - 30
CADMIUM           0 - 67
CHROMIUM          0 - 57
COPPER           33 - 80
LEAD             13 - 67
MERCURY          50 - 99
NICKEL           14 - 87
SELENIUM         50
SILVER           33 - 50
THALLIUM          0 - 13
ZINC             52 - 95
```

Indicator parameters, gross organic indicators, heavy metals, and abiotic measurements all indicated that surface water quality was not degraded by the apparent release of leachate impacted ground water to the surrounding surface water. Concentrations of contaminants decreased or were at ambient levels. Gross organic and abiotic indicators showed deterioration of water quality as the distance from the landfill increased. This indicates that other sources, such as surface run-off, the drainage canal, etc. appear to be major degrading influences on the surface water.

OBSERVATIONS

1. Data from field and laboratory studies indicate that the landfill appears to have no adverse impact on the ambient environment.

2. The peat layer is reducing the nonpersistent and persistent organics to acceptable levels. The only organic compound noted to be migrating through the peat was methylene chloride in the range of 200 ug/l, one order of

magnitude below the aquatic criteria limit for marine and freshwater organisms. This could be laboratory contamination

3. Sampling of ambient waters around the site indicates that there is little or no leachate related contamination from the site. There is substantial contamination due to runoff and drainage canal discharge to the surrounding water from other sources. This was determined by the decrease in surface water quality moving away from the landfill along sampling transects radiating from the site periphery.

4. Parameters which should be monitored for detection of possible impacts are:

TOC (soluble) — Note any discharge of problem organics (greater than 80 mg/l)

This will detect any significant organic compound(s) above ambient background levels.

BOD_5 (soluble) — Note any release of degradable organics (greater than 30 mg/l)

This provides an indication of biological activity.

pH (S.U.) — Note any acidic or caustic release (less than 6.0 or greater than 8.0)

This will detect significant change in hydrogen ion content, indicative of changes in abiotic conditions.

Sp. Conductivity — Note any change in salt levels (greater than 15,000 umhos/cm)

This will monitor changes in salinity, especially chlorides due to seasonal rainfall fluctuations.

ORP — Note any change in the nature of biologics (greater than +/- 300 mv).

A significant change indicates a change in anaerobic activity.

FOLLOW-UP PROGRAM

Results of this study were reviewed with appropriate state agency personnel. The unique nature of the site environment was apparent, and the merits of different alternative landfill construction designs and sitewide environmental measures were discussed in relation to their impact on the site environment.

The decision by BFI to incorporate positive containment for the site by means of sheet piling, compacted clay, and a dike surrounding the site, and leachate collection and disposal, was prompted by a desire to provide absolute assurance that no adverse environmental impact from the site would be possible. The assurance offered by natural sorption and attenuation provides outstanding back-up for the site over the long term, and would probably be adequate to meet current and proposed standards without additional extensive engineering and construction measures being taken.

CONCLUSIONS

Chemical and biochemical activities associated with landfills, landfill leachate, soils, and groundwater pose many questions for chemical engineers and scientists to answer. Applied research, field studies, and laboratory research are needed to fully understand the reactions of chemicals and microorganisms in aqueous solutions (leachate and groundwater) in and with the soil and rock.

This study demonstrated that the particular soils and microorganisms at this site can effectively remove some metals and organic compounds, and provided valuable input for the engineering design of site containment and closure facilities.

INDEX

A
Akron, Ohio 14
Ash testing 36

B
Biological 81

C
Characterization 115
Cleaning 127
Coal 94
Cofiring 94
Columbus, Ohio 25
Combustible materials 44
Combustuion 127
Consequences 107

D
Dade County 21
Disposal 107

E
Economica 1
Emission(s) 36, 94
Energy system 14
Environmental 107, 115

F
Flue gas 127
Fluidized bed 44
France 71
Full-scale 94

G
Gas 127

H
Holland 75

I
Incineration 44
Incinerator 115
Italian 54

J
Japan 44

M
Mass-burn 30
Materials 1, 44

Metro 21
MSW 44, 54, 107, 127
Municipal solid waste 54, 107, 127

O
Ogden Martin 30
Options k1, 81

P
Pelletized 94
Pelletizing 63
Plant 36, 75
Practices 107
Process 54, 71
Processing 75

R
RDF 14, 21, 25, 30, 36, 81, 94, 115
RDF/coal 94
Recovery 1, 21, 44
Recycle 14
Reduction 71
Refuse derived fuel 14, 21, 25, 30, 36, 81, 94, 115
Resources 21

S
Solid waste 54, 71, 107
State-of-the-art 1, 127
Studies 94
Success story 14
Sweden 63
System(s) 1, 14, 81

T
Technologies 127
Technology 115
Thermal 81

U
Utilization 63

V
Valorga 71

W
Waste 54, 63, 71, 75, 107
World 75

SYMPOSIUM SERIES

ADSORPTION

- 96 Developments in Physical Adsorption
- 117 Adsorption Technology
- 219 Recent Advances in Adsorption and Ion Exchange
- 230 Adsorption and Ion Exchange—'83
- 233 Adsorption and Ion Exchange—Progress and Future Prospects
- 242 Adsorption and Ion Exchange: Recent Developments
- 264 Adsorption and Ion Exchange: Fundaments and Applications

AEROSPACE

- 33 Rocket and Missile Technology
- 52 Chemical Engineering Techniques in Aerospace

BIOENGINEERING

- 69 Bioengineering and Food Processing
- 84 The Artificial Kidney
- 86 Bioengineering ... Food
- 93 Engineering of Unconventional Protein Production
- 99 Mass Transfer in Biological Systems
- 108 Food and Bioengineering—Fundamental and Industrial Aspects
- 114 Advances in Bioengineering
- 163 Water Removal Processes: Drying and Concentration of Foods and Other Materials
- 172 Food, pharmaceutical and bioengineering—1976/77
- 181 Biochemical Engineering Renewable Sources of Energy and Chemical Foodstocks

CRYOGENICS

- 224 Cryogenic Processes and Equipment 1982
- 251 Cryogenic Properties, Processes and Applications 1986

CRYSTALLIZATION

- 110 Factors Influencing Size Distribution
- 193 Design Control and Analysis of Crystallization Processes
- 215 Nucleation, Growth and Impurity Effects in Crystallization Process Engineering
- 240 Advances in Crystallization From Solutions
- 253 Fundametnal Aspects of Crystallization and Precipitation Processes

DRAG REDUCTION

- 11 Drag Reduction
- 130 Drag Reduction in Polymer Solutions

ENERGY

Conversion and Transfer

- 5 Heat Transfer Atlantic City
- 57 Heat Transfer, Boston
- 59 Heat Transfer Cleveland
- 75 Energy Conversion Systems
- 79 Heat Transfer with Phase Change
- 87 Advances in Cryogenic Heat Transfer
- 113 Convective and Interfacial Heat Transfer
- 118 Heat Transfer—Tulsa
- 119 Commercial Power Generation
- 138 Heat Transfer—Research and Design
- 162 Energy and Resource Recover from Industrial and Municipal Solid Wastes
- 174 Heat Transfer, Research and Application
- 189 Heat Transfer—San Diego 1979
- 202 Transport with Chemical Reactions
- 208 Heat Transfer—Milwaukee 1981
- 216 Processing of Energy and Metallic Minerals
- 225 Heat Transfer—Seattle 1983
- 236 Heat Transfer—Niagra Falls 1984
- 245 Heat Transfer—Denver 1985
- 257 Heat Transfer—Pittsburgh 1987
- 263 Heat Transfer—Houston 1988

Nuclear Engineering

- 53 Part XIII
- 56 Part XIV
- 95 Part XX
- 104 Part XXI
- 106 Part XXII
- 119 Commercial Power Generation
- 168 Heat Transfer in Thermonuclear Power Systems
- 169 Developments in Uranium Enrichment
- 191 Nuclear Engineering Questions Power Reprocessing, Waste, Decontamination Fusion
- 221 Recent Developments in Uranium Enrichment

ENVIRONMENT

- 78 Water Reuse
- 97 Water—1969
- 115 Important Chemical Reactions in Air Pollution Control
- 122 Chemical Engineering Applications of Solid Waste Treatment
- 124 Water—1971
- 126 Air Pollution and its Control
- 133 Forest Products and the Environment
- 137 Recent Advances in Air Pollution Control
- 139 Advances In Processing and Utilization of Forest Products
- 144 Water—1974: I. Industrial Wastewater Treatment
- 145 Water—1974: II. Municipal Wastewater Treatment
- 146 Forest Product Residuals
- 147 Air: I. Pollution Control and Clean Energy
- 148 Air: II. Control of NO_{xx} and SO_x Emissions
- 149 Trace Contaminants in the Environment
- 151 Water—1975
- 156 Air Pollution Control and Clean Energy
- 157 New Horizons for the Chemical Engineer in Pulp and Paper Technology
- 165 Dispersion and Control of Atmospheric Emissions, New-Energy-Source Pollution Potential
- 170 Intermaterials Competition in the Management of Shrinking Resources
- 171 What the Filterman Needs to Know About Filtration
- 175 Control and Dispersion of Air Pollutants: Emphasis on NO_x and Particulate Emissions
- 177 Energy and Environmental Concerns in the Forest Products Industry
- 184 Advances in the Utilization and Processing of Forest Products
- 188 Control of Emissions from Stationary Combustion Sources Pollutant Detection and Behavior in the Atmosphere
- 195 The Role of Chemical Engineering in Utilizing the Nation's Forest Resources
- 196 Implications of the Clean Air Amendments of 1977 and of Energy Considerations for Air Pollution Control
- 198 Fundamentals and Applications of Solar Energy
- 200 New Process Alternatives in the Forest Products Industries
- 201 Emission Control from Stationary Power Sources: Technical, Economic and Environmental Assessments
- 207 The Use and Processing of Renewable Resources—Chemical Engineering Challenge of the Future
- 209 Water—1980
- 210 Fundamentals and Applications of Solar Energy II
- 211 Research Trends in Air Pollution Control: Scrubbing, Hot Gas Clean-up, Sampling and Analysis
- 213 Three Mile Island Cleanup
- 223 Advances in Production of Forest Products
- 232 Applications of Chemical Engineering in the Forest Products Industry
- 239 The Impact of Energy and Environmental Concerns on Chemical Engineering in the Forest Products Industry
- 243 Separation of Heavy Metals and Other Trace Contaminants
- 246 Advances in Process Analysis and Development in the Forest Products Industries.

FLUIDIZATION

- 101 Fundamental Processes in Fluidized Beds
- 105 Fluidization Fundamentals and Application
- 116 Fluidization: Fundamental Studies Solid-Fluid Reactions, and Applications
- 176 Fluidization Application to Coal Conversion Processes
- 205 Recent Advances in Fluidization and Fluid-Particle Systems
- 234 Fluidization and Fluid Particle Systems. Theories and Applications
- 241 Fluidization and Fluid Particle Systems: Recent Advances
- 255 New Developments in Fluidization and Fluid-Particle Systems
- 262 Fluidization Engineering: Fundamentals and Applications

HISTORY OF CHEMICAL ENGINEERING

100 The History of Penicillin Production
235 Diamond Jubilee Historical/Review Volume

ION EXCHANGE

79 Adsorption and Ion Exchange Separations
219 Recent Advances in Adsorption and Ion Exchange
230 Adsorption and Ion Exchange—'83
233 Adsorption and Ion Exchange—Progress and Future Prospects
259 Recent Progress in Adsorption and Ion Exchange

KINETICS

25 Reaction Kinetics and Unit Operations
73 Kinetics and Catalysis

MINERALS

15 Mineral Engineering Techniques
85 Fossil Hydrocarbon and Mineral Processing
173 Fundamental Aspects of Hydrometallurgical Processes
180 Spinning Wire from Molten Metals
216 Processing of Energy and Metallic Minerals

PETROCHEMICALS

49 Polymer Processing
127 Declining Domestic Reserves—Effect on Petroleum and Petrochemical Industry
135 The Petroleum/Petrochemical Industry and the Ecological Challenge
142 Optimum Use of World Petroleum
212 Interfacial Phenomena in Enhanced Oil Recovery

PETROLEUM PROCESSING

103 C_4 Hydrocarbon Production and Distribution
127 Declining Domestic Reserves—Effect on Petroleum and Petrochemical Industry
135 The Petroleum Petrochemical Industry and the Ecological Challenge
155 Oil Shale and Tar Sands
226 Underground Coal Gasification: The State of the Art

PHASE EQUILIBRIA

2 Pittsburgh and Houston
6 Collected Research Papers
88 Phase Equilibria and Gas Mixtures Properties

PROCESS DYNAMICS

36 Process Dynamics and Control
46 Process Systems Engineering
55 Process Control and Applied Mathematics
214 Selected Topics on Computer-Aided Process Design and Analysis

SEPARATION

120 Recent Advances in Separation Techniques
192 Recent Advances in Separation Techniques—II
250 Recent Advances in Separation Techniques—III

SONICS

109 Sonochemical Engineering

MISCELLANEOUS

48 Chemical Engineering Reviews
70 Small-Scale Equipment for Chemical Engineering Laboratories
112 Engineering, Chemistry, and Use of Plasma Reactors
125 Vacuum Technology at Low Temperatures
143 Standardization of Catalyst Test Methods
182 Biorheology
183 The Modern Undergraduate Laboratory Innovative Techniques
185 Electro Organic Synthesis Technology
186 Plasma Chemical Processing
187 Chronic Replacement of Kidney Function
194 Hazardous Chemical—Spills and Waterborne Transportation
203 A Review of AIChE's Design Institute for Physical Property Data (DIPPR) and Worldwide Affiliated Activities
204 Tutorial Lectures in Electrochemical Engineering and Technology
206 Controlled Release Systems
217 New Composite Materials and Technology
220 Uncertainty Analysis for Engineers
228 Problem Solving
229 Tutorial Lectures in Electrochemical Engineering and Technology—II
231 Data Base Implementation and Application
237 Awareness of Information Sources
238 New Developments in Liquid-Liquid Extractors: Selected Papers From ISEC '83
244 Experimental Results from the Design Institute for Physical Property Data. I: Phase Equilibria
247 Chemical Engineering Data Sources
248 Industrial Membrane Processes
249 Measurement of High Temperatures in Furnaces and Processes
252 Thin Liquid Film Phenomena
254 Electrochemical Engineering Applications
256 Experimental Results From the Design Institute for Physical Property Data: Phase Equilibria and Pure Component Properties
258 Fiber Optics: Processing and Applications
261 New Membrane Materials and Processes for Separation

MONOGRAPH SERIES

3 The Manufacture of Nitric Acid by the Oxidation of Ammonia—The DuPont Pressure Process by Thomas H. Chilton
4 Experiences and Experiments with Process Dynamics by Joel O. Hougen
5 Present, Past, and Future Property Estimation Techniques by Robert C. Reid
6 Catalysts and Reactors by James Wei
7 The Calculated Loss-of-Coolant Accident by L.J. Ybarrondo, C.W. Solbrig, H.S. Isbin
8 Understanding and Conceiving Chemical Process by C. Judson King
9 Ecosystem Technology: Theory and Practice by Aaron J. Teller
10 Fundamentals of Fire and Explosion by Daniel R. Stull
11 Lumps, Models and Kinetics in Practice by Vern W. Weekman, Jr.
12 Lectures in Atmospheric Chemistry by John H. Seinfeld
13 Advanced Process Engineering by James R. Fair
14 Synfuels from Coal by Bernard S. Lee
15 Computer Modeling of Chemical Processes by J.D. Seader
16 "High-Tech" Materials by Sheldon Isakoff